MW01227349

# YOUR DEFENSES AGAINST THE CORONAVIRUS

JORGE LABORDA

# Your defenses against the coronavirus

## A brief introduction to the immune system

## Jorge Laborda

TITLE:
Your defenses against the coronavirus

AUTHOR:
Jorge Laborda



EDITION:
Jorge Laborda

LAYOUT:
Jorge Laborda

COVER DESIGN:
Jorge Laborda
Top right image: Novel Coronavirus SARS-CoV-2
This transmission electron microscope image shows SARS-CoV-2, the virus that causes COVID-19, isolated from a patient in the U.S. Virus particles are shown emerging from the surface of cells cultured in the lab. The spikes on the outer edge of the virus particles give coronaviruses their name, crown-like. Credit: NIAID-RML

PRINTING:
Lulu, USA



ISBN: 978-1-71682-346-6

*To Rosa*

*To the memory of my grand father*

*Daniel Fernández-Frechín*

## About the Author

Jorge Laborda is Professor of Biochemistry and Molecular Biology in the Faculty of Pharmacy at the University of Castilla-La Mancha, Spain. Among his scientific contributions, it is worth mentioning the discoveries that reveal the existence of two genes regulating the activity of Notch receptors, one of the most important for the control of cell growth and differentiation and for the functioning of the immune system.

During his work at the FDA, from 1991 to 1999, he was responsible for the evaluation of numerous projects on new anticancer therapies based on monoclonal antibodies. From November 2003 to May 2004, he was appointed as a Leading National Expert at the European Commission, where he worked on the management and promotion of the area of Synthetic Biology. In April 2004, he was elected Dean of the UCLM School of Medicine. From June 2007 to June 2011, he held the position of Councilor for Science, Technology and Consumer Affairs at Albacete City Council. Among his contributions in that capacity is the creation of the Promenade of the Planets, a scale reproduction of the Solar System: https://es.wikipedia.org/wiki/Paseo_de_los_Planetas.

Among the extensive contributions in the area of scientific popularization he was PI of eight popularization projects financed by the FECYT (Spanish Foundation for Science and Technology) for the popularization of the scientific activities of the UCLM through the program *Talking to Scientists* (http://cienciaes.com/entrevistas/). He has authored 20 books on scientific subjects. Twelve of these correspond to compilations of the more than one thousand popular science articles published in the newspaper *La Tribuna de Albacete*, newspapers of the Promecal group, and *El País*, and available in the blog *Quilo de Ciencia* (https://jorlab.blogspot.com). He has participated numerous times as a popularizer in *Vanguardia de la Ciencia* and *Hablando con Científicos* radio broadcast and podcasts programs. It also produces the podcast *Quilo de Ciencia*. He has also been the initiator and director of the podcast *Ciencia Fresca*. Finally, he was one of the initial promoters of the Spanish Science in the Parliament initiative, in which he participated as one of the six national experts to develop it.

# Table of Contents

# 1.- INTRODUCTION

There are numerous popular science books on the most diverse topics, but they are not abundant on Immunology. The paucity of popular science books on the immune system may be due to a variety of causes, among which I suspect the main one is that it is a difficult subject to explain so that people with basic biology skills can understand it. My experience in learning and teaching Immunology to Medicine and Pharmacy students at the University of Castilla – La Mancha indicates this. For this reason, I consider it a serious challenge, and a fascinating adventure, to try to explain the bases of how the immune system works in the simplest possible way, stimulating interested readers to learn more or refresh their knowledge, and also to health science undergraduate students to immerse little by little from the surface of this matter until reaching some of its most fascinating depths.

Undoubtedly, there seems to be a high interest in the "body's defenses", the name by which the immune system is popularly known. Maintaining the defenses in good condition is one of the goals of eating an adequate diet every day. There are even food products on the market that are advertised for their supposed role in maintaining the body's defenses healthy, rather than for their flavor qualities or nutritional value. It is equally true that vaccines are another issue of concern. How is a harmless vaccine made in such a way that the immune system "believes" nevertheless that it is a dangerous microorganism that needs to be eradicated? Does this manipulation of the immune system pose any danger? If so, what is its importance and what is the probability that it will materialize? We can try to answer these questions only if we know better some of the surprising processes by which the immune system functions, and the dynamic interaction between those and the microorganisms that try to survive by counteracting the effective methods used to annihilate them.

Although the defenses are interesting, it is no less true that they are a mysterious entity, of which it is only generally known that they serve to fight infections. However, new discoveries indicate that the body's defenses are essential to keep cancer at bay, and even essential for our ability to learn and remember and to maintain a good mood and not get depressed. Body's defenses in poor condition not only increase our risk of infection, but also increase the risk of other diseases and can even affect our mental balance. In addition, failures in the functioning of the defenses that lead to mistake our own cells for foreign microorganisms generate a variety of so-called autoimmune diseases, which include type 1 diabetes mellitus, multiple sclerosis, systemic lupus erythematosus, or rheumatoid arthritis, among the most frequent and known.

Probably, the interest in the body's defenses and vaccines has been tragically increased by the appearance of the SARS-CoV-2 virus pandemic, a new virus that belongs to the coronavirus family and that for that reason it is known as "the coronavirus". Infection with this virus causes the COVID-19 disease. Taking advantage of this interest, I must confess that I have hurried to publish this book so that it reaches as many readers as possible and as soon as possible. After explaining the main functional bases of the body's defenses, I will briefly address the origin of this pandemic, how the coronavirus infects us, and why some people do not suffer symptoms, others suffer them only mildly, and still others die from the disease caused by this terrible virus.

So, let's get into the fascinating world of the body's defenses. We are going, little by little, to know their protagonists, how they communicate with each other to coordinate their activity against enemies, how they learn to distinguish self from non-self, and how the body's cells manifest their identity, a continuous process which is essential so that the defense cells literally spare their life. We will try to compare the mechanisms of the body's defenses with known aspects of ordinary life and with more or less familiar systems, such as mask dances, which aim to mask the identity of those who participate in them, or the army, which is organized, in part, in a similar manner as the immune system is, since, as this does, its mission is to defend us from enemies and rebels.

To begin to understand the complex world of the immune system, we should treat it as if it were a photograph or a painting. This is important, because photographs and drawings are only understood when we look at them as a whole. It is not possible to capture the whole of an image focusing only on one of its corners, or on a lateral part. Likewise, an image would be difficult to grasp if we could only see a small portion of it every day, that is, being able to capture only a small proportion of its details each time we look at it, without having access to the globality of what is represented in it until we have contemplated it many times.

The above is one of the main difficulties that, in my opinion, needs to be confronted to get to a correct understanding of the immune system. Being this like an image, if we could see it globally "from above" as flying over it with a drone, we would understand it more easily and quickly. We cannot do that. We must unravel the image of the immune system little by little, as if it were a puzzle. We have no choice but to examine the pieces that compose it and, with patience, find out where and how they fit in the final image.

Obviously, there are simple puzzles and complex puzzles, depending on the number and size of the pieces that make them up. The same image can be broken down into a hundred large pieces or a thousand small ones. Fortunately, although the immune system puzzle contains thousands of pieces, these are grouped into larger pieces, which we can start using to compose it without having to separate the smaller pieces that compose them. It is as if it were a puzzle that can be solved, firstly, by beginners using the large pieces and, later, by experts who can already use the small pieces that make up the large ones, by knowing the shape of these and where they fit. Since we have no choice but to form a puzzle to understand the immune system, we are obviously going to start with the big-piece puzzle, a relatively simple puzzle. Once this puzzle is formed, we can dedicate ourselves to analyzing the smallest pieces in more detail and studying how they fit together to compose each large piece, until we are satisfied with our level of understanding. This second stage of solving the puzzle, which I will try to develop in the second half of the book, will depend on the motivation and interest of each reader and whether he or she wants to go deeper into this topic.

An important advice that I ask you to consider is that, in addition to being patient with the construction of the puzzle, you must build it at least twice. The functioning of the immune system from the first attack to the final victory is like an action novel: many things happen, and many characters and events are involved in the story. As with good stories, sometimes it is necessary and pleasant to read them two or even three times to extract all the juice, to really understand the motivations of the protagonists and the reasons for their actions. The same is true of the immune system, its characters and its motivations: it is convenient to read the story again to understand in depth the mechanisms and the reasons for its development. In short, dear reader, dear student, you do not need a lot of patience and dedication. However, I promise you that the patience and tenacity I ask of you will be worth it, because understanding immunology will also help you to understand the daily battles of life and the general functioning of systems, including the system of which you are part: the society.

I would love it if you decide to embark on this adventure with me. I assure you that, after completing this exciting journey along our defenses, you will be amazed forever with one of the most extraordinary acquisitions of Nature throughout evolution: your immune system.

# 2.- Defense strategies

## 2.1.- Damn it, we're surrounded!

Let's start with an important clarification: the immune system doesn't work only when we suffer an infection and fall ill, it works any time of day and night and, when it works correctly, it prevents us from getting infections and other diseases. The reason why the immune system is always working, as other organs or systems of the organism do, is that we are surrounded by potential enemies on all surfaces of our body; enemies that are in the air that we breathe, in the liquids we drink and in the food that we eat; enemies that are also on the surfaces of the body and try to penetrate our organism and often they succeed; enemies that once they have penetrated it is necessary to eliminate completely and without any mercy, or they will end up with our life. The immune system normally responds to all these threats in a very effective way. This response is called **the immune response,** which encompasses the set of actions that the immune system undertakes to defend us against the threats of the many enemies that seek to end with our lives. It is not an exaggeration; it's the harsh reality.

Body's epithelial surfaces are the first barrier of defense against the ongoing threat of enemy invasion. These surfaces are not limited to the outer skin, but also include what we might call "internal skins," such as those that line the intestine, lung, or secreting ducts of the genitourinary tract. Our bodies are like those ancient fortresses whose walls served to prevent enemy invasion. In our case, however, the enemy lives on the wall surface, lurking and waiting for any damage in it to penetrate the fortress. Numerous species of bacteria live on our skin, let alone on the epithelial surface of the lungs, excretory system, and intestine, that is, on any outer or inner surface of the body that is in contact with the external environment through any opening. It is estimated that the number of bacteria attached to our epithelial surfaces is greater than ten times the number of cells in our body. That is the magnitude of the potential enemies that haunt us at every moment, many of whom also possess fearsome counter-defensive mechanisms.

In our case, however, the walls are not formed by inert stones, but are alive, and act to keep enemies at bay, preventing them from penetrating. To defend ourselves, in addition to forming a layer, generally impenetrable to bacteria and other microorganisms while not damaged, in addition to repairing this layer quickly if damaged, cells involved in the formation of epithelial surfaces, such as the so-called keratinocytes (producers of the keratin protein of the skin, hair, horns, nails and hooves), and some immune system cells found in the skin, such as **macrophages,** produce various types of proteins. Several of them serve the mission of controlling the number of bacteria that can adhere to the epithelial surfaces, to minimize the likelihood that they could penetrate them. These include **mucins**, proteins containing high amounts of carbohydrates, making them sticky for bacteria. Mucins are fundamental components of the **mucus,** an adhesive and viscous liquid, secreted by the body's internal epithelial surfaces. These, in addition to having cells specialized in mucus production, also have cells that participate in the generation of movements that make it flow, which prevents bacteria from sticking to those surfaces and gets them to simply "swim" over them, attached to the mucus. For example, the intestine, thanks to the peristaltic movements necessary to circulate food, also circulates the mucus that it secretes, and that comes out as part of the stool, thus dragging out numerous intestinal bacteria. The surface of the lungs has cells with microscopic hairs called cilia, which are continuously in motion to flow the mucus secreted over it. The importance of producing mucus of adequate quality to defend against bacterial infections is evident in the disease called **cystic fibrosis**. This illness is characterized by the production of a mucus too thick and dehydrated, due to a defect in a gene necessary for its correct production. This thick mucus is not able to flow normally through the epithelial surfaces of the lung, giving to the bacteria greater opportunities to penetrate them, and leading these patients to suffer from recurrent lung infections caused by bacteria.

In addition to these physical procedures to prevent bacterial penetration, epithelial surfaces have numerous chemical antibacterial defense mechanisms. They involve the so-called antimicrobial peptides and proteins, as well as enzymes that digest certain components of bacterial walls. The bacterial wall is not a simple membrane, as is the case in eukaryotic cells. It is a molecular lattice that lines the bacteria outside its internal lipid membrane and gives it rigidity and protection against the entry of undesirable substances, including the entry of too much water from the outside, which would eventually inflate and make the bacteria explode.

The chemical defense mechanisms consist mainly of enzymes, including two: **lysozyme** and **phospholipase $A_2$**, secreted with tears, mucus and saliva. Lysozyme is an enzyme that leads to lysis, that is, the breakdown of bacteria and to their death. It acts by digestion of certain carbohydrates that form the fabric of the cell wall of, above all, Gram-positive bacteria.

What are Gram-positive bacteria? They are simply the type of bacteria that are dyed violet-blue in the Gram stain test. Let's make a brief parenthesis to explain it. Gram staining is due to Danish bacteriologist Christian Gram (1853-1938), who developed this staining technique in 1884. This consists mainly of the use of a dye called **crystal violet** and a discoloration procedure. The exposure of bacteria to crystal violet, along with other iodine-containing chemicals, causes crystal violet to penetrate inside bacteria and stain them all, both Gram-positive, and Gram-negative bacteria. A subsequent discoloration procedure is unable to cause Gram-positive bacteria to lose the acquired coloration; however, it is lost by Gram-negative bacteria. Thus, Gram-positive bacteria are those that do not discolor after staining, and Gram-negative bacteria are those that do.

The difference in behavior in the Gram's staining and discoloration procedure is partly due to the difference in the chemical structure of the bacterial wall. This difference in structure makes Gram-positive bacteria more susceptible to the action of lysozyme than Gram-negative bacteria. However, lysozyme can digest the wall of both types of bacteria, leading to water ingress through the bacterial lipid membrane and to the rupture (lysis) of bacteria.

Lysozyme is produced by cells present on various types of epithelial surfaces, and by cells called phagocytes, of which we will later talk about, and secreted to the outside. It is found mainly in tears, which thus prevent bacterial eye infections, in saliva, breast milk and mucus. Lysozyme is also secreted into the intestine by cells specialized in the production of this and numerous antibacterial substances, which are located at the base of so-called small intestine crypts. These cells are called **Paneth cells**.

Saliva, tears and milk also contain **phospholipase $A_2$**. In addition, Paneth cells produce and secrete this enzyme into the gut. Phospholipases are enzymes that degrade phospholipids, which are the main lipid components of bacterial membranes. By destroying phospholipids, phospholipase $A_2$ destroys the bacterial membrane and kills the bacteria. This may be somewhat surprising, but we must always keep in mind the idea that, in the case of cells, the fundamental units of life, what separates the living world from the non-living world is, in all cases, only one layer formed by two lipid molecules. This lipid membrane, or lipid bilayer, is the one that allows the imbalance of ions and substances necessary to maintain the energy flow of life. Lipid membranes work in this regard as hydraulic dams do, hindering the balance of the water level that would prevent energy from being obtained. Membranes maintain imbalances, also called gradients, on either side of them, imbalances that make life possible. The rupture of the membrane involves the breakage of the dam, the immediate collapse of the imbalance and, consequently, death.

Another group of antibacterial molecules secreted by epithelial cells and by the phagocytes present in them and in the tissues of the body are called **antimicrobial peptides**. Peptides are short fragments of proteins, formed only by the binding of a few amino acids to each other. Their small size and the chemical nature of the amino acids that form them make it possible for them to be inserted into lipid membranes and destabilize them, leading to the death of bacteria and also to the neutralization of some viruses that are coated by lipid membranes stolen from the cells to which they parasitize. There are three main classes of these antimicrobial peptides: **defensins**, **cathelicidins** and **histatins**, although they all act on lipid membranes, preventing them from maintaining the imbalance between both sides that makes life possible.

These substances pose an immediate risk to microorganisms penetrating the epithelial walls. It is as if when attacking a castle, all those unfortunate soldiers who tried to get to the top of its walls were poisoned by substances produced by the stones and, if they survived the poisoning, they were seriously weakened. There is no doubt that poisonous walls would have been a good defense strategy in castles of the Middle Ages, if anyone had thought about it and could have developed the technology to make them real. Well, organisms such as us have "invented" those walls along evolution and partly because of them we can survive every day.

| Defense type | Skin | Intestin | Lungs | Eyes, nose, mouth |
|---|---|---|---|---|
| Physical | Barrier formed by occluding junctions between epithelial cells | | | |
| | | Longitudinal liquid flux | Movement of cilia | Tears, nasal cilia |
| Chemical | Epidermis and dermis fat | Acidity | Mucus (adhesive) | Enzymes of tears (lysozyme) |
| | | Enzymes (pepsin), mucus | | |
| | Antibacterial peptides: defensins, cathelicidins, histatins | | | Phospholipase A2 (saliva) |
| Microbiological | Normal microbiota | | | |

***Table I – Defense mechanisms of epithelial barriers***

## 2.2.- Holes in the wall

However, what is not invented yet are totally impenetrable walls. Even Nature, the greatest inventor of all time, has not succeeded in this endeavor. What happens then, when the wall is damaged and can be penetrated by enemies attached to it surface, many of which are able to survive despite the antibiotic substances it produces? This is when, in addition to the fascinating wound healing mechanisms, the extraordinary mechanisms of the immune system are put in place.

Damage to the wall can be caused by a wound, or by a chemical aggression (e.g. too much alcohol intake, which can damage the intestinal wall), depending on the type of wall in question and its location in the body. Damaged walls can be penetrated by bacteria, viruses or fungi that may live on them, on the surface of the skin or intestine, for instance.

What happens if an enemy penetration occurs? This is when things start to get interesting and the immune system needs to be activated immediately to stop the invasion. Whether it gets activated sooner or later depends on several factors. One of them is luck, on which it depends that immune system cells find soon enough the enemy that has penetrated and react against it in a short time. Another factor is the state of proteins and cells in the immune system: whether there is an adequate amount of them, whether the cells are well fed, etc. Let's take a closer look at this situation.

Suppose that when sewing a button, we prick ourselves with the needle. The epithelial barrier of, for example, our index finger of the left hand has been damaged. A drop of blood begins to appear on our skin. Blood flow and activation of blood clotting are factors that make it difficult for bacteria on the skin to penetrate the wound easily. Blood flowing out of the wound expels bacteria outwards. Rapid clotting attempts to plug the huge hole (considering the size difference between the tip of a needle and a bacteria) that the needle has pierced into the epithelial wall and nearby blood vessels. However, some bacteria have already penetrated by the puncture, pushed by the needle, and managed to adhere to the inside of the skin and try to establish an enemy colony in the dermis or even deeper, depending on the magnitude of the puncture.

In the latter case, several things can happen. The first is that bacteria will be bathed by the body's internal fluids, particularly those present in the dermis. These already contain antibacterial substances and enzymes that will attack and try to prevent bacteria from establishing an enemy colony in which to reproduce. Body fluids also contain immune system proteins that form the so-called **complement system,** which we will talk about later in more detail **(section 2.3)**. This system is activated by detecting certain wall components from all types of bacteria and leads to the coating of bacteria with proteins that facilitate their capture and destruction by certain cells of the immune system. The complement also leads to the formation of pores in the bacterial membrane that place in contact both sides of it, which breaks the well-known imbalance needed for life and leads to the death of bacteria.

In any case, if only a small number of bacteria have penetrated, they are usually eliminated without us realizing it. However, despite being surrounded by toxic substances for them, some bacteria can survive, as they have fascinating mechanisms for evading the immune system **(section 7)**. If this happens, it is also possible that bacteria surviving to the action of toxic substances, luckily for us, are immediately found by one or a few immune system cells, which I like to call **sentinel cells.** The sentinel cells (later we will give more details about their features) are located inside the skin, of which they are part, where they reside waiting for potential enemies that have been able to penetrate the epithelial wall. Keep in mind that daily activity can damage this wall in several ways: rubbing with objects, bumps, cuts, etc. It is necessary to maintain a large sentry cell detachment on the inner surfaces of the epithelial walls to control the multiple invasion attempts that can occur each day.

If the penetration of bacteria has happened, fortunately, in a place of the skin where one or more sentinel cells are located, they run across the enemy and act immediately, joining forces with the complement system and other antibacterial substances. Your mission is, in general, to try to capture this enemy and eliminate it by digesting it inside. In addition, each sentinel cell that encounters an enemy gives the alarm (by producing and secreting to its environment certain molecules that we will see later), with the aim of attracting to the place where the invasion attempt has occurred more sentinel cells and other cells that I call "soldier" cells, in particular **neutrophils** and **monocytes**, which will become **macrophages,** which, along with neutrophils, will chase down and capture bacteria. Once captured, these cells "eat" the bacteria. For that reason, they are called, in scientific language, **phagocytes**, a word that derives from two Greek words: *phagein,* meaning 'to eat', and the suffix *-cyto,* which means ´cell'.

In addition to phagocyting and destroying bacteria, macrophages and neutrophils can produce oxidizing substances that are highly toxic to microorganisms and can also cause damage to our own tissues. These substances include **nitric oxide**, **superoxide anion** (negatively charged) and **hydrogen peroxide**. Other toxic substances can be generated from this, such as **hydroxyl anion** (OH-) and **hypochlorite** (OCl-), an anion with antiseptic properties that is also found in bleach. These substances are produced in an enzymatic process called **respiratory burst**, since oxygen is needed to generate them. The respiratory burst does not usually occur, however, at the beginning of the infection, but when the infection has advanced and it is necessary to use the most powerful methods available to eliminate it. Later, we will see how the immune system decides under which conditions the respiratory explosion should or should not be triggered.

If, together, the first sentinel cells and phagocytes that arrive capture and digest, or kill with their enzymes and antimicrobial peptides, all the bacteria that have penetrated the wound, before they have time to reproduce, nothing bad has happened and the sentinel cells and phagocytes resume their functions of surveillance and patrol with the "cellular and molecular feeling" of a job well done, a job that no one has known about and that no one will thank them for, even though they have managed to save the lives of all the cells of the organism. Anonymous heroes never properly recognized, these poor sentinel cells and those poor phagocytes do not know that they are.

However, it is possible that, after the needle has pierced our skin, thus allowing the penetration of the bacteria, no sentry cells will find them. The bacteria that have penetrated and are able to resist the action of the antimicrobial substances always present in the tissues, are, for this reason, momentarily at least, in a bacterial paradise: an area at an ideal temperature, humid and overflowing with nutrients and where the toxic substances present on the skin cannot do them enough harm. Stimulated by so much sudden abundance, the bacteria begin to celebrate in the only way they know how: by starting to reproduce like crazy, and this in spite of the fact that they do not enjoy sex. Bacteria thus establish what is called a source of infection, i.e. a place in the body from which they threaten to infect everything and can even collaborate with each other to achieve this.

This deserves a short digression. Since when we get a wound, we get blood, we may believe that the bacteria that have entered the wound will pass into the blood. Although it is possible that some of them may pass, even though the blood comes out and expels the bacteria to the outside, these bacteria will not be part of the source of the infection, as they will be carried by the bloodstream into the body. Normally these bacteria will be eliminated by passing through **the spleen**, which is the organ specialized in cleaning the blood of microorganisms, especially during childhood, in addition to participating in the general functions of the immune system. The bacteria in a source of infection are not found in the blood, but in our tissues or organs, i.e. located outside the blood vessels and capillaries of the circulatory system, which form a complex and intricate system of branched tubes through which the blood circulates. This does not exclude that, in some occasions, the bacterial infection is so important that, in fact, it ends up reaching the blood and spreading through the body. In that case, the subsequent widespread immune response throughout the body can cause the so-called **septic shock**, which is potentially fatal. We will discuss this possibility in more detail later, but for the time being we will limit ourselves to localized infections at specific points in the body, such as the point where we have pricked our finger with the needle.

In the favorable conditions of the interior of our organism, the bacteria reproduce every twenty or thirty minutes, which they do by growing and dividing in two, a process that is called cellular fission. Division by cellular fission results in an increase in the bacterial population in a geometric progression. From an initial bacterium, in about thirty minutes at most, we have two; in an hour four; in two hours sixteen, and so on. In general, we are not aware of the power of geometric progressions. Let's take a little break to calculate how the bacteria would grow if they had unlimited resources and nothing stopped them from growing. The mass of the Earth is just under six thousand trillion trillion grams ($6.0 \times 10^{27}$ grams). A typical bacterium in our intestine, *Escherichia coli*, weighs about 95 picograms ($95 \times 10^{-12}$ grams), with one gram being equal to one trillion picograms. How long would it take for a bacterium to reach the mass of our planet, reproducing every thirty minutes? The calculations that I have made, and that I have had to repeat several times to believe them, indicate that in just a little more than three days a simple bacterium would have reproduced until its descendants surpassed the mass of the planet Earth. Only eight more hours would be needed for the bacteria to reach the mass of the Sun. This is the enormous power of exponential growth.

Obviously, bacteria are never in ideal growing conditions. Not all bacteria that reproduce give viable offspring, so the geometric progression is not as rapid as the one outlined above. Nor do they have unlimited resources, because the organism has several mechanisms to limit them. The growth capacity of bacteria is, however, phenomenal, and it is especially so when they are in the right conditions inside our body, having penetrated an epithelial barrier. For this reason, if the bacteria introduced into our body after being punctured by the needle are able to survive in the face of toxic substances and do not encounter any sentinel cells, they can generate a source of infection within a few hours, in which thousands of bacteria will be found. If a thousand bacteria survive initially, in half an hour we will have about two thousand; in an hour, about four thousand, and so on. Under these conditions, when a sentinel cell finally encounters one or more of these bacteria and detects them, even if it sounds the alarm, the other sentinel cells and "soldier cells" that come in may not be able to cope with the ever-growing bacteria. Bacteria that manage to establish a source of infection threaten to invade the entire body. Before the discovery of antibiotics, only the cells and molecules of the immune system could stop them, which they have done, whenever possible, for hundreds of millions of years during the evolutionary history of animals. However, despite antibiotics, infectious diseases remain the leading cause of death in the world.

It is worth making another parenthesis here. When we say that bacteria try to invade the whole organism it is obviously not that they have the conscious intention of doing so. Bacteria respond to unconscious molecular mechanisms which, if not controlled and restrained, will in fact lead to the invasion of the whole organism and to the death of the animal, and to the death of most of the bacteria that have invaded it. Unrestrained invasion is not the best strategy for the survival of the invading organism itself, but it is nevertheless where bacterial growth would lead if not restrained by the immune system.

A fierce battle takes place at the source of the infection. On the one hand, the bacteria are being phagocytized by the sentinel cells and by the phagocytes that come to the call of those cells that first encountered the bacteria. The cells that are arriving in response to the initial alarm signal reinforce the alarm call and get more and more phagocytes, the most effective cells for the search and capture of bacteria. It is important to understand from this moment that the alarm signals are molecules, and signals are transmitted by molecular modifications. When we talk about signals in the world of cells we are always talking, in fact, about molecules and processes in which some molecules act on others to transmit information from one part of the cell to another, particularly from the cell membrane to the nucleus. Thus, phagocytic cells have molecules on their cell membrane that can detect certain molecules produced by bacterial metabolism, molecules that bacteria cannot avoid producing if they want to continue living. These molecules attract phagocytes to the bacteria. In fact, phagocytes can produce so-called pseudo-podia, or false feet, and move with them in pursuit of the bacteria until they are captured and ingested. Bacteria, however, are not easily captured. Some are equipped with a layer of specific carbohydrate molecules that makes them elusive for phagocytes and protects bacteria from being captured by them. Therefore, if the number of bacteria is too large to allow all of them to be destroyed, the most that phagocytes can do is to contain the rate of expansion of the infection and help set in motion other more effective mechanisms of antibacterial attack. These mechanisms depend on the activation of specialized cells of the immune system: **lymphocytes**, which we will discuss later. For the moment, let's remember this and get on with the fight.

## 2.3.- The complement system

Fortunately, as we have said, sentinel cells and phagocytes are not alone. They are not the only means of defense found on the inside of the damaged epithelial wall. Every tissue in the body is bathed in a liquid. This fluid is similar to blood plasma, although it does not normally contain blood cells, except precisely when they leave the blood to attack a source of infection, or to patrol the body.

The fluid that bathes tissues and organs contains several molecules that act against bacteria. First, it contains certain molecules produced by the liver in response to infection (so-called acute phase proteins, which we will discuss later). Second, it contains a molecular defense system called the complement system. This molecular system specializes in detecting certain molecules found in bacteria, or bacteria and other microorganisms coated with antibodies (very interesting defense molecules that we will also discuss in detail later). The complement therefore detects, directly or indirectly, this latter thanks to antibodies, molecules of the bacterial enemies that try to infect us. As if this were not enough, although when it detects an infection it is strongly activated, the complement system is always spontaneously active anyway, whether it has detected bacteria or not, in case any bacteria, virus, or other microorganisms that may have penetrated to the body fluids go unnoticed by the macrophages and start a source of infection or start infecting our own cells by not being eliminated in time. The complement system is always in a state of alert against the enemies.

This system is made up of twenty-five proteins that are in an inactive state, but which are activated in what constitute **three pathways of biochemical reactions**, initially independent, and then converging at a common point, from which the biochemical mechanisms are identical. The three independent pathways, activated in three different but complementary ways to fight bacteria, lead from the activation of initial proteins to the activation of intermediate proteins, already common to all three payhways, and finally to the activation of a final protein complex, equally common to all three cascades. The three cascades are called the **classical pathway**, the **lectin pathway** (lectins are proteins that bind to carbohydrates) and the **alternative pathway**. The activation of the initial proteins occurs by proteolysis, i.e. by lysis or fragmentation of inactive precursor proteins which, when lysed enzymatically, generate the active components.

The three complement pathways are not activated in response to the same triggers. In fact, the alternative pathway, so named because it was discovered as an alternative to the others, is, however, the most important one. This is the one that is always spontaneously activated at a basal level even in the absence of infection. This indicates that an activation of the complement system "just in case" is important to keep potential microorganisms that can penetrate the epithelial barriers at bay. The need for this continued activation is also indicative of the persistent threat of infection that we face, since, throughout evolution, only those individuals who were able to develop it appear to have survived. The lectin pathway is activated when certain acute phase proteins (discussed in the next section) bind to carbohydrates present on the surface of the bacteria. Finally, the classical pathway is activated when the first protein of this pathway, called **C1q**, binds to antibodies attached to the surface of microorganisms, to the surface of some bacteria directly, to **C-reactive protein**, or to **mannose-binding lectin (section 2.4)**, an acute-phase protein. The activation of all these complement pathways fulfils various missions, which we will try to explain below.

The most important step in the action of complement is the lysis and activation of an intermediary protein. This protein, called **complement factor C3**, is the point of convergence of the three activation pathways, which until now have been activated and progressing independently of each other. This means that any of the three proteolytic cascades leads to this same point, which is the crucial point in the process of complement activation. This point involves the generation of an active enzyme called **C3 convertase** which, as its name suggests, converts the C3 protein from its inactive to its active form. The C3 protein, when activated by proteolysis through the action of this enzyme, binds through a covalent bond to the surface of the bacteria and leaves them marked for destruction by various mechanisms. This is important because many species of bacteria defend themselves from being captured and destroyed by phagocytes by coating themselves with layers of molecules that make them elusive to these cells and prevent their capture. However, when protein C3, after being activated, binds to the surface of the bacteria and coats them, which the bacteria cannot prevent in any way, the bacteria cannot escape from the phagocytes and are efficiently phagocyted by them. This process of coating the bacteria and antigens in general to promote their phagocytosis is called **opsonization**. The reason for the increased efficiency of phagocytosis is that phagocytes possess on their surface receptor molecules for the activated C3 protein bound to microorganisms. By binding to these molecules with several of their C3 receptors at the same time, which can only happen if several C3 molecules bound to the surface of a microorganism are close to each other, the bacteria are captured and the phagocytes activate the phagocytosis process, introduce the bacteria inside and digest it by means of digestive enzymes. For this reason, activation of the C3 protein in enough quantity is essential for infection control. People who, for one reason or another, lack adequate levels of this protein in the blood, or lack the appropriate control mechanisms to allow its proper activation, are susceptible to bacterial infections.

Complement activation progresses beyond this intermediate stage with the activation of a final molecular complex. The activation of the final proteins of the three pathways leads to 18 units of the last component, the so-called **C9 protein**, spontaneously assembling together to form tiny pores that pierce the surface of the bacteria and are deadly. The pores are indeed tiny, because their diameter is about 10,000 times smaller than that of a human hair.

Despite its small size, this is enough to exert its deadly effect. The pores in the membranes of any cell cause its death because the cell membrane, consisting of only two layers of molecules of a fatty nature, is the barrier that separates life inside the cell from non-life outside. The formation of pores in the membrane brings both worlds, the living and the non-living, into contact, and when that happens the non-living world always prevails and causes death. Bacteria perforated by this complement protein complex die because the outer fluid enters the pores, the inner bacterium being a more concentrated solution than the outer medium, and ends up swelling the bacterium and causing it to shatter. In addition, as the internal medium of the bacteria meets the external medium, the pores break the ion imbalance between the two sides of the bacterial membrane, an imbalance that is fundamental to bacteria for obtaining energy from the metabolism of the nutrients.

A serious problem with this state of affairs is that complement activation, especially activation by the alternative pathway, the most important one, does not discriminate between bacteria and our own cells. Pores can form in both. Fortunately, our cells, if they are healthy, have proteins on their membrane that stop the formation of pores if they begin to form. This prevents our cells from dying by the same process that complement kills bacteria.

Although the structure of the pores has been determined through studies with electron microscopy and other techniques, until recently the dynamic process of their formation had not been observed. This has been achieved by using a microscopic technique called rapid atomic force microscopy, which works by obtaining information not through light, but through touch, by sliding a very small needle over the surface of what is to be examined to detect changes in its texture. In this case, what the scientists examine is an artificial bacterial surface on which they activate the complement so that it forms the pores. This study allowed scientists to find out a hitherto unknown fact. When the last activated protein of the complement, as we have said, protein C9, must be inserted into the membrane to start forming the pore together with 17 of its partners, the process stops for a brief moment. This brief pause is vital. During this moment, if the pore is being formed in one of our cells, it allows time to stop its formation thanks to the membrane proteins that stop this process. This brief pause does not, however, affect the ability to form pores in bacteria, which lack the proteins capable of preventing their formation.

Thanks to these studies, we see more clearly the marvelous processes and their tunings that have been generated during the evolution of animals to keep us alive, preventing deadly bacterial infections and, at the same time, preventing these processes from damaging us too much. The process of complement activation is finely tuned in time so that our cells can defend themselves from its harmful effects, but not the bacteria, which will perish phagocyted or perforated, with no remedy for them.

## 2.4.- Acute-phase proteins

Complement proteins are continuously synthesized by the liver and secreted into the bloodstream from where they also diffuse into tissues. Complement proteins are therefore always available in case they are needed to help overcome an infection attempt.

However, this is not the only way in which the liver helps to defeat infections. When sentinel cells and phagocytes detect bacteria at the site of infection, these cells produce and secrete into the blood numerous proteins that serve to send an alarm signal to other phagocytes and attract them to the site of infection. These proteins also serve to alert the liver that an infection attempt is taking place. These proteins secreted by phagocytes and sentinel cells, and generally by the various cells of the immune system, are called the generic name of **cytokines**.

Cytokines transmit information about the type of microorganism or parasite that is trying to infect or penetrate the body and serve several important functions. One of them is to raise body temperature, causing fever. Fever plays a significant role in accelerating the immune response. In addition to raising body temperature and other functions, the cytokines produced by phagocytes act on the liver, which detects the presence of the increased concentration of concrete cytokines and reacts by producing the so-called **acute phase**. In this phase, the liver increases the production and secretion of certain proteins and decreases the production of certain others, generating changes in the proteins of the blood plasma that aim to prevent the progression of infectious microorganisms. Among the proteins that increase their amount in plasma are **mannose binding lectin** (mannose is a carbohydrate similar to glucose, often found on the surface of bacteria), which is capable of activating complement by the lectin pathway, and **C-reactive protein**, which binds certain lipids on the membranes of some bacteria and is able to activate complement by the classical pathway, leading to opsonization and subsequent phagocytosis of the microorganisms, or to their destruction by the formation of pores in their membranes, as explained above **(section 2.3)**.

Other acute phase proteins are fundamental in another very important aspect of defense: **the control of the nutrients** that the microorganisms need for their reproduction. We mentioned earlier that the bacteria that have been able to penetrate the skin after the needle puncture are in a heavenly place, at an ideal temperature and with an abundance of nutrients. Well, one way to stop the growth of microorganisms is to make that paradise a little less generous by controlling the access of microorganisms to a nutrient resource that is indispensable to them: **iron**. Iron is absolutely necessary for bacterial growth, and if this element cannot be captured by bacteria in sufficient quantity, even though they have other nutrients in abundance, they cannot reproduce. Generating an iron deficiency in bacteria is an effective method of preventing their growth, however, this is not easy.

As we know, iron is certainly abundant in the body. Red blood cells contain enormous amounts of hemoglobin, the protein that carries oxygen from the lungs to the rest of the tissues and organs, which is charged with four iron atoms for each molecule. Hemoglobin can be released from red blood cells into the blood, and, therefore, it is an important iron source for invading microorganisms. Some of these also produce toxins that attack the red blood cells and break them down, a process called **beta-hemolysis**, or affect the hemoglobin so that it releases the bound iron even without breaking down the red blood cells, a process called **alpha-hemolysis**, name used for this process although in this case it does not produce, as we say, rupture or lysis of the erythrocytes. Alpha-hemolysis is, however, sufficient to get the iron out of the red blood cells and into the blood plasma. The microorganisms capable of generating any type of hemolysis are powerful pathogens, since they can cause anemia and compromise the transport of oxygen to the tissues.

Fortunately, several acute-phase proteins produced by the liver serve the purpose of sequestering iron from the blood and body fluids and preventing it from being captured by bacteria, which makes it difficult for them to grow. Two of the most important acute phase proteins for iron control are **ferritin** and **haptoglobin**. Ferritin captures iron present in blood and tissue fluids and facilitates its incorporation into cells. In this way, the amount of iron available to the infectious microorganisms present in these fluids decreases. The ferritin gene is activated in response to infections, so more of this protein is produced, more iron is captured, and more is incorporated into the cells.

Haptoglobin plays a similar role to ferritin, although instead of binding to iron directly, haptoglobin binds strongly to hemoglobin that could have leaked into blood plasma from red blood cells. This binding allows the hemoglobin in the blood plasma to be captured by the cells of the spleen and thus removed from the bloodstream and body fluids.

In addition to these, there are other acute-phase proteins whose production increases in response to infection. There are also blood proteins whose production is decreased by the liver, since their normal levels are not strictly necessary and, in case of infection, amino acids, the basic molecules that form all proteins, are preferably used to produce the acute-phase proteins that must increase to thus defend us from the uncontrolled growth of microorganisms.

So far, we have analyzed the mechanisms that the immune system uses as a first line of defense against a wide variety of microorganisms. These mechanisms are part of what is called **innate immunity**. However, this immunity is not always able to eradicate the invaders. If they continue to progress, innate immunity will set in motion more expeditious immune mechanisms, which are characteristic of so-called **adaptive immunity**. To understand how this immunity, which adapts to each microorganism, is set in motion, it is necessary to return to the center of the battle, to our source of infection.

## 2.5.- Adapting to the inner enemy

Let us remember that, within the source of infection, bacteria are reproducing at great speed. More sentinel cells and phagocytes are coming to the site, which have detected and responded to the molecules produced by the sentinel cells and phagocytes that initially detected the danger. However, if the bacteria have had time to grow before being discovered, when they reach the source of infection these cells encounter an already large population of bacteria in continuous reproduction. Under these conditions, phagocytes are not able to eradicate them, even with the help of the complement system and acute-phase proteins. More expeditious mechanisms are needed. Specialized squads need to be trained and recruited to fight an enemy that has become established in the organism and threatens to destroy it. As we have mentioned, these special and very effective squads are made up of cells, called lymphocytes, which do not come to the site of infection until they have learned to identify the enemy and have been activated and armed in the right way to eradicate it. Let's see how these special forces are put into action, without which many infections could not be eradicated.

When we accidentally wound ourselves, it is advisable to wash the wound abundantly with water and soap. However, throughout our evolution and the evolution of animals, washing wounds was not an obvious option, although they were usually licked. Saliva contains a factor (called factor III) that stimulates blood clotting and also stimulates the activity of some immune cells. In addition, as we have said, saliva also contains antiviral and bactericidal compounds, such as lysozyme, which destroys the wall of many bacteria, causing water to invade, swell and break them, killing them. Therefore, licking the wounds is beneficial. However, saliva also contains bacteria that nest in the mouth, which could also cause infections.

In any case, washing the outside of the wounds, with soap and water or with saliva, is not the only wash that takes place. It may be surprising that the wounds are also subjected to internal washing, by means of the plasma fluid, the same fluid that forms blood. How and why does this internal washing take place? To understand this, we will have to go into a subject that is a staple in the activity of the immune system: **cellular communication**. Cells "speak" to each other with a language that is made up not of words, but of molecules secreted into the extracellular medium, which are captured, not by ears, but by other molecules that act as detectors and receptors of the former and which are found on the surface of the cells. Without this communication between cells, the internal washing of wounds, and practically all the processes of the immune system, would be impossible.

Let us remember that the sentinel cells have detected a group of bacteria that has established itself in a source of infection. We have said that these cells send out molecular alarm signals that attract more sentinel cells and phagocytes to this source. Particularly, neutrophils, the main phagocytes, are the first to come to the wound and join the fight. This pilgrimage of cells to the source of infection has even been recorded on video. However, before they can reach the source of infection, a serious problem must be solved: these cells must leave the blood by crossing the blood vessels towards the tissues, in search of the bacteria of the source of infection. How do they achieve this?

Well, they achieve this thanks to a set of surprising mechanisms that would be impossible without the communication of sentinel cells and phagocytes with cells that are not normally considered as immune system cells. These cells are none other than the cells on the internal surface of the blood vessels, the so-called **endothelial cells**, since they are part of the endothelium.

The endothelium is, again, another wall. It is like another epithelial surface, formed by cells strongly attached to each other, which, thanks to this strong adhesion, keep the blood and the cells it contains inside the blood vessels. Although the endothelium allows the exchange of gases, nutrients and waste, and the blood plasma may come out in small quantities from time to time, most of it is normally well trapped within the circulatory system. The situation changes dramatically, however, when an infection attempt occurs.

By detecting the first bacteria at the source of infection, the sentinel cells generate and secrete specific cytokines to the outside, which in this case, in addition to acting on the liver, can also be detected by the endothelial cells near the point of infection. These molecules serve as an activating signal to the endothelial cells, which, in response to these cytokines, especially to a very important one, called **TNF-α** (Tumor Necrosis Factor alpha), will relax the strength of the bond between them. This relaxation of their binding strength allows the vascular diameter to increase, the blood circulation to slow down at that point, and the plasma fluid, with the complement components and the acute-phase proteins, to flow out between the endothelial cells into the wound and begin to wash it out, while the complement acts, causing opsonization and perforation of the bacteria. At the same time, the activated endothelial cells produce and place on their surface certain molecules that are sticky for phagocytes. When phagocytes are pushed by the blood flow through that area near the point of infection, many remain attached. This adhesion is facilitated by the reduced speed of the blood flow that has occurred. This area of the endothelium has become sticky for phagocytes, but not for other blood cells, such as red blood cells and platelets, which do not adhere to these sites. This adhesion of phagocytes to the endothelium allows the implementation of molecular mechanisms that enable weakly attached phagocytes to adhere much more strongly to endothelial cells, first, and to leave the blood vessel by passing between two endothelial cells, which have decreased their binding strength, second. This process of passing between two endothelial cells is called **extravasation**. In this way, the relaxation of the binding force between the endothelial cells, induced by the cytokines produced by the sentinel cells, allows the surroundings of the wound, where the infection is being generated, to be flooded with plasma fluid. This, together with the fluid that is already bathing the tissues, forms the **lymph**, a fluid that will transport, through the **lymphatic vessels** (similar to blood vessels), bacterial debris and even whole bacteria from the wound to the **lymph nodes**, which are the organs that receive the lymph and are responsible for the activation of the lymphocytes, whose name literally means 'lymph cells'. At the same time, phagocytes which have come out of the blood in the process of extravasation are also accumulating in the source of the infection. These are going to phagocytize the bacteria and secrete more cytokines. Some of these phagocytes, in particular some macrophages and some sentinel cells, the so-called **dendritic cells**, which we will discuss later, will also be carried by the lymph into the lymph nodes.

Another effect of the cytokines released by the sentinel cells at the site of infection is that they can induce blood clotting in the blood capillaries near the site. Coagulation is induced by the cytokines reaching the blood and being carried away by the blood, so it occurs mainly downstream from the site of infection. This coagulation does not therefore make it difficult for fluids and cells to leave the blood at the site of the infection, but it does make it difficult for bacteria that can enter the blood from the site of infection to spread to the rest of the body through the bloodstream. Coagulation facilitates the accumulation of lymph and cells near the site of infection by blocking or at least slowing down the flow of blood away from that site. However, it clearly causes damage to the tissues downstream of the source of infection by making normal blood circulation difficult, thus preventing adequate oxygen and nutrient supply to those tissues. This is already a first "**collateral damage**" caused by the defensive action of the immune system. Like all wars, the one in which the immune system fights microorganisms always causes more or less collateral damage to the body itself.

In some cases, the collateral damage can even cause the death of the organism if the bacterial infections are not controlled. We have briefly mentioned above the so-called **septic shock**, a particularly severe case of sepsis, caused by the passage of bacteria into the blood. If the initial infection is not eliminated and the bacteria grow faster than the immune system can handle, which can happen in certain situations where the defenses are not in good shape, for example as a result of surgery or malnutrition, the infectious bacteria or their molecular components can spread through the blood or lymph to the whole body. This situation leads to all the sentinel cells located on the various epithelial surfaces and organs of the body reacting to what they assume is a local infection, which is happening only where they are, but which is actually happening in all parts of the body at the same time. Dendritic cells and macrophages in the body react normally, releasing cytokines, including TNF-$\alpha$. As a result, the endothelial processes to fight microorganisms are triggered, especially the relaxation of the blood vessel walls and the coagulation of small capillaries. As we have already said, relaxation means that the bonding strength between the endothelial cells diminishes and the endothelium becomes less impermeable to the blood. This leads to a significant outflow of plasma fluid from the blood into the tissues, which causes **edema**. Edema leads to a dramatic drop in blood pressure that makes it difficult to deliver oxygen to important organs, such as the liver, kidney, brain, or heart itself, even though in that situation the heart speeds up the frequency of its beating in an attempt to provide sufficient blood flow to the organs that need it. If the situation is not corrected by urgent medical treatment, with antibiotics, intravenous fluids, injection of red blood cells and anti-inflammatory and vasoconstrictor drugs, sepsis can lead to death.

Fortunately, sepsis is an unlikely complication in the case of infections caused by everyday wounds, so let's go back to the endothelium of the blood vessels near the source of infection. What are the molecules that, first, allow the adhesion and then the extravasation of phagocytes and lymphocytes from the circulatory system to the tissues and organs and what properties do they have? There are several of these molecules and each of them plays a fundamental role in the extravasation process. Firstly, we have certain carbohydrates and the proteins that interact with them, which are called **selectins**. Selectins are part of a large family of proteins, called **lectins**, whose function is to interact and adhere to carbohydrates. The cells of the endothelium produce two selectins which they place on their surface: **E-selectin** and **P-selectin**. These selectins appear on the surface of endothelial cells a short time after being activated by the cytokine TNF-$\alpha$. Both selectins bind to a carbohydrate present on the surface of lymphocytes, monocytes and neutrophils, called **sialil-Lewis$^{X}$**.

The main difference in the function of both selectins is their operating time. P-selectin is stored in granules of endothelial cells, called Weibel-Palade bodies. The signal sent by the cytokine TNF-α lead to the release of the granule contents to the cell surface in just a few minutes, so that the endothelial cells become adhesive very soon after the release of TNF-α. This cytokine simultaneously induces the activity of the gene for E-selectin synthesis, which, although produced rapidly, cannot appear immediately on the surface of endothelial cells because it must be produced first. However, only two hours after being stimulated by TNF-α endothelial cells already express on their surface an abundant number of E-selectin molecules.

The sticky interaction between the sialil-Lewis$^X$ molecules of phagocytes and lymphocytes and the selectins of endothelial cells is essential to allow adhesion of monocytes, neutrophils and lymphocytes to the endothelium near the source of infection. The reason for this requires a detailed description of what is happening in the blood vessels. As almost everyone in the civilized world knows, they carry the blood that is being pumped by the heart. The blood contains an enormous number of platelets and erythrocytes, among which the other blood cells are packed together. Dragged by the rapid flow of blood, all these cells collide with each other and with the endothelial walls of the arteries and veins. It is important that when the cells collide they do not adhere to each other and, above all, they do not adhere to the endothelium, except if it is strictly necessary, or otherwise there would be blockages in the progression of the bloodstream that could cause serious problems. For this reason, the endothelium is normally a smooth surface and is not sticky to any blood cells. We can imagine it as smooth as a marble surface on which billiard balls, also completely smooth, slide at full speed. These balls are the erythrocytes, among which, from time to time some tennis balls appear, which in this simile play the role of monocytes, neutrophils and lymphocytes. These, like tennis balls, have some "little hairs" on their surface. These "hairs" are the sialil-Lewis$^X$ molecules. In spite of these "hairs", tennis balls also glide at high speed and unimpeded on the marble surface of the endothelium, unless the endothelium cells detect the presence of TNF-α, produced by the sentinel cells at the site of infection, in which case the endothelial cells begin to produce a kind of "molecular Velcro" on their surface. Although this Velcro will not affect the speed of the billiard balls, it will interact with the tennis ball hairs and slow them down. This "molecular Velcro" is formed by the molecules of E-selectin and P-selectin.

The leukocytes thus slowed down by the adhesion of their sialil-Lewis$^X$ with the endothelium selectins continue to be pushed by the blood flow, but since they are now stuck, albeit still weakly, on the surface of the endothelium, retained by the sticky interactions with the selectins, they now roll on its surface much more slowly than the cells that cannot interact with the endothelial selectins. This slow rolling provides the time needed to allow much stronger adhesive interactions to take place, which fix the cells to the endothelium and stop the rolling altogether.

These stronger adhesive interactions are those established between so-called **adhesion molecules**, present on endothelial cells, and **integrins**, present on monocytes, neutrophils and lymphocytes. The most important adhesion molecules in the endothelium are called **ICAM-1** and **ICAM-2**. ICAM stands for Intercellular Cell Adhesion Molecules. Adhesion molecules are made up of a single amino acid chain that has several **immunoglobulin domains** in its structure. Remember that a domain is a region of a protein chain that folds in space independently from the rest of the protein. We will study the importance of immunoglobulin domains later, when we talk about antibodies. For the moment, we will only mention that the presence of immunoglobulin domains in many proteins of the immune system is a frequently repeated theme. Therefore, all proteins with immunoglobulin domains have been included in one large family, in fact, in a superfamily of proteins, which is called the **immunoglobulin superfamily**. After being stimulated by TNF-α, endothelial cells near the sites of infection or parasite entry sites increase the amount of ICAM-1 and ICAM-2 molecules that appear on their surface. This allows passing monocytes, neutrophils and lymphocytes, carried by the blood circulation, to adhere strongly to them and to leave the circulatory system and enter the tissues, where they can join the battle against the enemy.

As noted above, the molecules of monocytes, neutrophils and lymphocytes that adhere strongly to the molecules of ICAM-1 and ICAM-2 and allow them to bind to the endothelium, despite the strong blood flow that tries to carry them, belong to another family of proteins: **the integrin family**. We see here that there are families of proteins with functions that depend on the interaction between one and the other. Thus, the adhesion proteins of the ICAM family, in general, have a partner to which they adhere that belongs to another specific family, in this case the integrin family, which is formed by two different protein chains, called and α and β, which act in combination like a pincer to adhere to the ICAM molecules. Proteins formed by two different chains are also very common in the immune system. These types of proteins are called **dimeric proteins**. As we shall see, dimeric molecules of the immune system are, in general, **heterodimeric**, because each of their two components is different from the other, which is what the *hetero* prefix means. In the case of integrins, there are several different α chains and several different β chains that combine with each other in various possible ways, giving rise to different integrins that can adhere to one or another adhesion molecule. In addition to ICAM-1 and ICAM-2, there are other molecules that exert different functions within the immune system, all of which are related to cell adhesion and communication.

The strong adhesion of leukocytes and lymphocytes to endothelial cells fixes them to the endothelium and completely stops their rolling on the endothelium surface, despite the strong blood flow that could detach them. This allows the establishment of further interactions between the endothelial cells and monocytes, neutrophils and lymphocytes, interactions that allow these cells to glide between two endothelial cells and pass through the endothelium to the other side. The immune cells then penetrate the basement membrane of the blood vessels with the help of enzymes that break down the proteins in the extracellular matrix of these vessels. Finally, once the blood vessel is completely traversed, they can now be directed to the focus of infection. This process is called **diapedesis**, or also **extravasation**.

However, the fact that the phagocytes leave the blood between the endothelial cells is not enough for them to reach the focus of infection and contribute to eradicate the bacteria there. The phagocytes that have left the blood must also know where to go, i.e. they must find out exactly where the bacteria against which they must fight are located. In this respect, phagocytes and, in general, all the cells of the immune system have a serious problem, just one more: **they are blind and deaf**. How can they find their way around once they have come out of the blood and are in a three-dimensional, dark, silent world with no sign of where to go?

Fortunately, although they are blind and deaf, phagocytes, and in general the cells of the immune system, possess an excellent sense of smell, or something like this sense. They are capable of "smelling" certain molecules and heading for the source of the "fragrance". The molecules that phagocytes and lymphocytes can “smell” are called **chemokines**. However, it is important to mention here a fundamental difference between the sense of smell and the ability to detect chemokines possessed by leukocytes and lymphocytes, which is that not all cells are capable of “smelling” all chemokines. Specific cells only have "smell" for specific chemokines and ignore all others. In this way, they do not make mistakes about where they should go. For example, neutrophils detect the chemokine called CXCL8, which is produced by macrophages that have detected microorganisms at the site of infection, while monocytes, which in addition to going to the site of infection must spread through tissues, where they will become sentinel macrophages even in the absence of infection, also detect the chemokine called CCL2. Thus, neutrophils never leave the tissues unless there is an infection, while monocytes do, thus forming the sentinel macrophage population of all tissues.

Let us make a brief parenthesis here to explain the origin of the words "cytokine" and "chemokine", which are obviously related. The first derives from the union of two words of Greek origin: *cyto* and *kine*. The second derives from the union of two other words: *chemo* and, again, *kine*. The prefix *cyto* is derived from the Greek word *kytos* and means 'cell'. On the other hand, the prefix *chemo* does not need to be explained much. It refers, of course, to a chemical, because chemokines are chemicals; in fact, they are usually proteins. The ending *kine* is somewhat more mysterious. It is derived from the Greek word *kinesis*, which means 'movement'. This word is part of others like "kinetics", or "kinematics", always related to movement. We can now deduce, if we have not already done so, that cytokines and chemokines are chemicals that will make the cells that detect them "move". In fact, the cytokines will activate many immune cells to perform defensive functions, and the chemokines will literally make them move in space towards the direction where there is a greater amount of these substances, which is usually either in the lymphatic organs or in the source of infection.

Thus, the first sentinel cells that encounter the bacteria at the site of infection produce cytokines that activate endothelial cells, which relax and become sticky to allow plasma fluid and immune cells to escape. At the same time, the sentinel cells that detect the bacteria make and release chemokines to the outside. These chemokines diffuse and generate a concentration gradient, that is, they spread from their site of origin in all directions, decreasing their molecular density as they move away from the site where they are produced, similar to how a drop of ink would diffuse in water, or an odor diffuses in the air. When they are detected by monocytes, neutrophils and lymphocytes, they leave the blood vessels first, and then begin to move in the direction of the sentinel cells that are producing the chemokines, which is none other than the source of infection. In this way, phagocytes and sentinel cells, although blind and deaf, possess a kind of sense of smell that guides them and enable them to find their way to join the hard battle for survival against the invading bacteria.

The outflow of fluid from the blood vessels and the outflow of phagocytes to the site of infection cause the volume of this site and its surroundings to increase. Therefore, an **inflammation** occurs, which means an increase in volume in that area. The cells that have been leaving the blood must navigate towards the source of infection through the cells of the infected tissue and through the so-called **extracellular matrix**, the set of proteins that hold the tissues together. Many chemokines bind weakly to component molecules of this extracellular matrix, thus forming a kind of pathway by which the cells can be directed towards the source of infection. Navigation and diffusion through the body's tissues is not easy and, to facilitate this, phagocytes and other cells near the site of infection produce and release enzymes called **metalloproteases** to the outside. Metalloproteases are enzymes that require certain metals for their activity, and partially digest the proteins of the infected tissues, thus making it easier and faster for the leukocytes to reach the bacteria. For this reason, the inflamed tissues appear less consistent and softer than normal. In addition, some of these proteases also attack the bacteria directly, trying to digest them.

Because what occurs in the first moments of the immune response to an enemy that has penetrated the epithelial walls is an inflammation, this part of the immune response (response to an enemy threat, in this case) is called **the inflammatory response** or simply **inflammation**. However, these inflammatory processes manage to partially liquefy the tissues, and in addition to facilitating the action of phagocytes, prevent bacteria from adhering to the tissues too easily, which can facilitate their capture and elimination. Still another function of the inflammatory response, as we have seen, is to induce blood coagulation in the local microcapillaries to avoid the dissemination of the microorganisms through it to the rest of the body as much as possible.

Bacteria and their debris can thus be carried away by the fluid that has been leaking out of the blood vessels, but that must be returned to the blood. This fluid, the lymph, is removed by the lymphatic system, which consists, as the most important components, of the lymphatic vessels and the **lymphatic organs**. It is in the latter that the sentinel cells and the debris of the dead microorganisms will arrive and where the cells specialized in fighting against the particular enemy that has established itself in the source of infection will be activated: the lymphocytes. Let us clarify that, initially, the lymphocytes do not come to the source of infection, since they have not yet been activated. Only after activation in the lymph nodes will some of these lymphocytes, those specialized in "battlefield" fighting, leave the lymph nodes and reach the source of infection.

Lymphatic vessels are blood-like vessels, although their walls are thinner and contain a kind of valve that allows the passage of fluid and the cells they carry in one direction only. This direction is from the periphery of the body to the lymphoid organs and from there to the blood. The connection to the blood circulation system occurs at the so-called **thoracic duct**, which is the largest lymphatic vessel, and the **left subclavian vein**, which is located at the base of the neck. At that point, the lymph (the fluid that has left the blood vessels and washed away the wound) plus the cells of the immune system that are floating inside are reintroduced into the blood. The energy for this continuous flow of lymph is provided by the heartbeat.

Before returning to the blood, however, the lymph must fulfill two very important missions. The first, as we have said, is to wash the wound. Let us remember that a few pages ago we were talking about an internal washing of the wounds. The accumulation of fluid near the source of infection, and the action of proteases, manages to drag with the lymph many bacteria and their debris. At the same time, numerous sentinel cells that have phagocytized the bacteria are also dragged with the lymph. These sentinel cells are of two types that we remember here: **dendritic cells and macrophages that were residing in the tissues**. Later, we will explain in more detail the defense missions carried out by these important cells. The second mission that the lymph must carry out is to transport the dendritic cells and the macrophages to the lymphoid organs so that they can present the enemy they have captured to the lymphocytes. This is where things start to get very interesting.

### 2.5.1.- Information and the immune system

Before we go on to describe the ferocity of the battle, it is worth taking a few moments to explain concepts that I believe are very important for understanding what happens when the immune system reacts to a threat of infection. One of them is the concept of information management that the immune system must develop.

It is obvious that the immune system must collect information about the enemies in the outside environment that threaten to invade the entire organism. What kind of microorganism is attacking me? Is it a bacterium, a virus, a fungus, a worm? Once it has gathered this information, the immune system needs to make decisions accordingly. These decisions involve putting in place the necessary mechanisms to deal with the threat in question: viruses, bacteria, etc.

Each type of microorganism needs several molecules to survive and maintain its way of life. These molecules cannot be replaced by others, since they are necessary for a vital function. Bacteria, for example, need specific molecules for their cell wall, and viruses have their own proteins and nucleic acids. Given their fundamental nature for the life of microorganisms, these molecules are common to many of them, but they are not found in our body or, in general, in animal organisms. The presence of these molecules in the body therefore carries the information that a microorganism is trying to invade it. Since these molecules are associated with microorganisms, they receive the generic scientific name of **Microorganism-Associated Molecular Patterns (MAMPs)**.

Over the course of evolution, the immune system of animals has acquired genes that produce receptor and detector proteins for these molecular patterns that function as signals of the presence of microorganisms. One of the most important families of MAMP receptor proteins is the **Toll-like receptor (TLR) family**.

There are thirteen TLRs (TLR-1 to TLR-13), although the human species only has the first ten. Each of these is specialized in detecting some component or repetitive molecular pattern characteristic of some type of microorganism, but absent in eukaryotic cells. For example, the receptors TLR-2 and TLR-4 detect wall components of different types of bacteria (including a very important one: **lipopolysaccharide -LPS-** of Gram-negative bacteria); TLR-5 detects a protein necessary for the functioning of bacterial flagella and TLR-9 detects foreign nucleic acids, thanks to the absence of certain chemical modifications in them, modifications that are found in our nucleic acids. However, some Toll-like receptors are also able to detect molecules that are characteristic of our own damaged cells and tissues, molecules that are generically called **DAMPs (Damage-Associated Molecular Patterns)**. On the other hand, the TLR-10 receptor seems to function as an inhibitor and acts to regulate the immune response so that it is not too amplified, which can cause excessive collateral damage.

TLRs are present on the outer or inner membranes of dendritic cells and macrophages, but also on other cells of the immune system or on some epithelial cells. When TLRs detect a given molecule from a type of microorganism, they transmit this signal into the cell. The transmission of the signal entails the activation of certain molecules that are found in the cellular cytoplasm in an inactive form and that are activated by the molecular changes taking place in the TLRs that have detected the foreign molecules. These changes allow the internal part of the receptor, located in the cytoplasm, to interact with the molecules that will transmit the signal provided by the detection of the microorganisms' molecules.

The activation of these signaling molecules in the cytoplasm invariably leads to one or more specific proteins travelling from the cytoplasm to the cell nucleus, where they act as **transcription factors** and set certain genes in motion. The most important transcription factor activated by the TLR receptors is the so-called NF-κB (Nuclear Factor kappa B), but this is not the only one, because each Toll-like receptor can activate a set of transcription factors that act coordinately together.

What are transcription factors? Well, they are proteins that act in the nucleus as triggers for the functioning of genes. Genes that are not active are not being transcribed, that is, their information is not being expressed. In order for the information of a gene to be used, for example, for the manufacture of a protein, this information must be copied from the DNA, where it is stored, to the messenger RNA, the only nucleic acid that can be used by the ribosomes for the synthesis of proteins according to this information. Transcription is the generation of a messenger RNA from the sequence of "letters" in the DNA. For transcription to take place, transcription factors must physically bind to certain sequences of "letters" that are normally located slightly above the "letters" containing the information to be used for protein synthesis. Without the binding of the transcription factors to those DNA sequences, the genes are normally switched-off, silenced. The binding of a transcription factor to a gene sets in motion the synthesis of its messenger RNA and makes it possible to translate it into protein, i.e. it leads to the information stored in the DNA being manifested, expressed, in the world inside the cell, and makes it possible for the cell to perform a function that it could not perform before the gene was set in motion. Thus, each type of cell in our organism including, of course, the immune system cells, has a set of transcription factors activated, which are those allowing the genes that maintain the cells' "personality" to be expressed. This is fundamental for the development of the function that each cell must carry out in coordination with the other cells of the organism.

Obviously, not all genes are the same, nor do all genes need to be activated or silenced in response to an external signal, such as the detection of a molecule from a microorganism. The genes that are to be set in motion by a particular external signal is a process finely regulated in each cell type. This regulation has several layers. Let's mention some of them.

A first layer is what kind of receptors capable of detecting external signals are present on the cell membrane. Not all cells have the same receptors and therefore not all detect the same external stimuli. In other words: according to their initial personality, each cell is equipped to detect only certain external signals.

A second layer of regulation is provided by the existence of different classes of receptors, which differ in terms of the transcription factors they will eventually activate. We have already mentioned that many of the TLRs activate the NF-κB transcription factor. Other receptors, however, activate other transcription factors.

Finally, a final layer of regulation is provided by the "letters" (the nucleotide sequence) that each gene possesses, and which allow transcription factors to bind or not and activate them. In this way, the activation of a given transcription factor only results in the activation of the genes to which it can bind, which are only those that have the sequence of "letters" that allows them to bind.

Thus, in the case of cells of the immune system, the information detected from the outside allows the cell to make a series of molecular decisions. The first of these decisions is to set in motion the genes that will allow it to react correctly and effectively to the type of threat in question. This reaction can lead to the implementation of defense mechanisms and the elimination of enemies, and to the implementation of mechanisms for transmitting to other cells the information detected. This transmission is carried out through the generation and secretion of cytokines (produced from the activated genes), which will in turn activate receptors present on other cells of the immune system, resulting in the activation of other transcription factors and other genes in those cells that will now enable them to perform actions that contribute to the overall defense.

#### 2.5.1.1.- Genes and cellular missions

I consider it necessary here to open again a small parenthesis to explain the concept of gene as I believe it is necessary to understand it in order to make clear the activation and the actions of immune system cells. Genetics is frequently in the media today, and DNA-based paternity tests, or DNA analyses of biological samples obtained at the crime scene, are frequent in real life and in works of fiction. Therefore, we may think of a gene as a piece of DNA and believe that genes are in DNA and are DNA. Well, this is not entirely true. If DNA is necessary to contain genetic information, a gene is more than just DNA, since the gene is nothing, nothing, unless it manifests itself in the real world. The gene only makes sense as a method of producing a piece of the cellular machinery, or a component of a cellular system, that enables cells to do something that without the functioning of the gene they could not do. For example, if cells need to phagocytize bacteria, they need tools and methods to capture, internalize and digest them. All of these processes depend on the functioning of certain genes that produce the pieces of the cellular machinery that enable these functions, genes that are active in the cells that phagocytize bacteria, but that are not functioning in the cells that do not perform this function, such as in neurons. It should be said here that the functioning of genes is called, in scientific language, **gene expression**. A gene that is expressed, that manifests itself in the real world, is simply a gene that is functioning at that moment. A gene that increases its expression is a gene that increases the rate of its functioning. A gene that does not work is a gene that has been silenced, that has lost its expression, and a gene that stops working in response to some molecular signal is said to have silenced its expression. Genes, by the way, do not have freedom of expression, since their expression is always controlled by various molecular events and by the expression of other genes.

Every cell in our body has a functioning set of genes, i.e. a set of genes that are being expressed, and it also has a non-functioning set of genes, i.e. a set of silenced genes. The set of genes that a cell has functioning is what enables it to perform the functions that it should carry out. For example, a kidney cell has a functioning set of genes that is different from the set of genes in a liver cell. This is the reason why both cells are different. Cell differentiation, in fact, is the process by which different daughter cells are generated from a parent stem cell. Although daughter cells have the same genes on their chromosomes, they do not have the same genes being expressed. The process of differentiation entails that of the set of genes that the stem cell has functioning, and that allow it to be a stem cell, some genes will be switched off to give rise to daughter cells. Furthermore, in these daughter cells, which are going to become different from the stem cell (and in many occasions also different from each other, since a stem cell can give rise to various types of daughter cells), some genes that were silenced in the stem cell will become activated and others that were activated are going to be silenced. After several division processes, at the end, the precursor stem cell generates differentiated daughter cells that will be able to perform different functions according to the set of genes that each one has now working. This is, in fact, the most important meaning of gene expression.

Gene activation and gene silencing are particularly important molecular processes for immune system cells. Although they must be prepared to detect the presence of an enemy, they do not always have all the weapons ready to fight it or to present it to other cells so that they can fight it. These weapons are produced or increased after the enemy has been detected, and this production, or its increase, depends on the increase in the level of expression of certain genes and on the decrease in the level of expression of others. This has its logic, because it is not convenient to let fully armed and killing cells swarm in the organism if it is not necessary. In addition to the danger they would pose, it would also be a waste in terms of energy to keep a large population of cells equipped and armed if not for the immediate struggle. It would be just as absurd to keep a large fraction of a country's population mobilized in the army, equipped with weapons and supplies, even if the threat of war is not imminent. Security and the economy are age-old concerns of the immune system, two problems that it has had to manage hundreds of millions of years before the first human army marched on the planet.

Let us now return to the source of infection.

### 2.5.2.- Introducing the enemy

As we have said, dendritic cells and macrophages are the two types of cells that first detect invading bacteria, with or without the help of complement. To do this, they have the TLR receptors, which we have discussed above. These receptors are actually detectors in the first place.

As we have also said, these receptors detect molecules typical of bacteria or viruses, such as double-stranded RNA, lipopolysaccharide (LPS) from the wall of some classes of bacteria and several other molecules distinctive of microorganisms.

When these molecules are detected, the most important result is that dendritic cells and macrophages are activated. This means that they set in motion mechanisms and processes against bacteria that were not activated before they were detected. One of these processes is phagocytosis, as we have already pointed out. Phagocytosis allows these cells to ingest bacteria and digest them, which is in itself an important means of fighting them. However, phagocytosis is not the only process by which dendritic cells, and other phagocytic cells, capture microorganisms. Another process is called **macropinocytosis**, by which dendritic cells ingest large quantities of extracellular fluid and the suspended particles it may contain, which may include microorganisms. In any way the microorganisms are ingested, they are digested inside the cells.

This digestion kills bacteria, but it is not a total digestion. Not all bacterial components are fully digested. Some of them are only partially digested and are going to be used to show the nature of the enemy to the lymphocytes and to activate them against that enemy. Lymphocytes possess so-called **antigen receptors** and many of them are permanently moving from one lymph node to another, transported by the circulation of blood and lymph, in search of sentinel cells that show on their surface an enemy molecule that fits into one of their receptors. Each lymphocyte has an antigen receptor for a different molecule, and since there are billions of different lymphocytes, each with its own receptor, they are all capable of detecting and reacting against virtually any enemy molecule. These receptors will be discussed at length later.

Dendritic cells and macrophages that have captured and digested some bacteria, as well as debris from death bacteria, and even some live bacteria (which will be captured and digested later), are carried with the lymph through the lymphatic vessels to the lymph nodes closest to the source of infection. As a result of the influx of lymph and cells from the swollen tissues, the lymph nodes near the point of infection may also swell. We have probably all noticed our swollen neck nodes at some point as a result of a throat infection. This node swelling is a clear sign that a secondary phase of the immune response is being set in motion: **the adaptive immune response**.

Before the lymphocytes can be activated, only **the innate immune response** is available. This type of immune response is orchestrated by cells that respond generically to a variety of enemies, i.e. a multitude of bacteria, fungi or viruses, and is made possible by the presence on the membrane of these cells of Toll-like receptors/detectors and other receptors for molecular components of microorganisms, as well as receptors for some chemokines and cytokines. However, if, despite the fact that more and more innate immune cells come to the source of infection and more and more lymphatic fluid with complement proteins accumulates in that source as well, the cells of the innate immune system cannot contain the invading microorganism, the lymphatic fluid, with many bacteria, activated sentinel cells, is drained by the lymphatic vessels and carried to the lymph nodes. It is within these nodes that the lymphocytes, the most effective cells against the microbial infection, will be activated. These cells are not going to fight all bacteria or viruses, in general, as the cells of the innate immune system do, but they are going to "learn" the nature of the enemy they must face and generate a particular response against it. This response will be extremely effective against the microorganism in question, but it will be completely ineffective against a different microorganism that may infect us at the same time or in the immediate future. That is why the adaptive immune response is said to be specific: it is directed only against one species of microorganism, but not against others.

How are lymphocytes activated against the enemy? Well, this depends on the type of lymphocyte involved, as there are two main types: **B cells and T cells**. The reason they are called this way is because

B-lymphocytes are generated in the bone marrow (bone, hence B) and T lymphocytes, although in their early stages of maturation they are also generated in the bone marrow, develop in their final phase in an organ called **the thymus** (hence T), which is located in the chest, in front of the heart and behind the breastbone. These two organs, the bone marrow and the thymus, are therefore called **primary lymphoid organs**, since they are responsible for generating the primary cells of the immune system. In contrast, the lymph nodes in which these cells are to be activated are called **secondary lymphoid organs**.

### 2.5.3.- Activation of B lymphocytes

Let's go into activation of B-lymphocytes first. These lymphocytes are activated if they detect any molecule, either as bacterial or viral debris, or present in living bacteria or viruses, although sometimes they can also detect molecules from the body itself, which can cause autoimmune diseases. To detect molecules that come, as we will see, from practically any chemical entity, B-lymphocytes are equipped with receptors, i.e. molecules on the surface of their cell membrane, of virtually infinite diversity. Later on, we will see how B-lymphocytes manage to generate receptor molecules capable of detecting virtually any chemical structure in the world, but let us now focus on their activation.

There are billions of B lymphocytes in our bodies, each capable of detecting a particular molecule, usually presented by a bacterium or a virus. This molecule is usually called by the generic name of **antigen**. B lymphocytes do not know which molecule their receptor will detect. However, if a B cell receptor is activated by an antigen molecule to which the receptor can bind, three things happen, all of which depend on changes in the expression of given genes.

The first thing that happens is that the B lymphocyte that has detected the foreign molecule is stimulated to divide and form copies of itself, a phase of the activation process that is called **clonal expansion**. From a single lymphocyte capable of detecting a specific substance, thousands or more lymphocytes will be generated, which will be initially identical copies of the original, i.e. clones of it, and will therefore be able to detect the same substance. This phase of clonal expansion requires several days to complete. This is the reason why the adaptive humoral response (and, as we will see later, the cellular response carried out by the activation of T lymphocytes) takes several days to be generated. During those initial days of infection, the immune system must rely only on the defenses provided by the innate immunity.

The second thing that happens is that the B lymphocytes that have detected a specific antigen and have activated and divided will produce large quantities of antibodies that will be secreted to the exterior and transported by the blood. The secretion of antibodies is the main function of B lymphocytes. The antibodies are proteins very similar to the receptor molecule that the original B lymphocyte has on its membrane, used to detect the antigen against which it has reacted and it has been stimulated to multiply. In fact, antibodies are **molecules identical to that receptor**, except that they lack the part that allows them to bind to the lymphocyte membrane. Lacking this part of the molecule, the antibodies are secreted to the outside environment. Once in the blood or extracellular fluids, the antibodies will bind to antigens present in the bacteria and other microorganisms and make their life difficult in various ways. One of these ways is called **neutralization**. As this name indicates, antibodies neutralize the enemy, preventing it from exerting its activity or from harming us. Neutralization is achieved when antibodies bind to molecules that the microorganisms need to infect the cells and penetrate inside them, or they bind to molecules that microorganisms produce to exert a damage that benefits them, as, for example, the bacterial toxins. The binding of an antibody to a toxin prevents it from binding to a cell membrane receptor, which could lead to the penetration and killing of the cells, thus leading to the release of nutrients that the cells keep inside and whose release would favor bacterial growth. Likewise, an antibody that binds to a protein that a virus or bacterium needs to adhere to the cells and infect them will neutralize that virus or bacterium, preventing it from being able to infect the cells. Another way that antibodies help defeat the enemy is by coating the bacteria and making it easier for them to be phagocytosed. Remember that this coating is called opsonization. The binding of antibodies to the surface of the bacteria makes it easier for the bacteria to be detected by the phagocytes through specific antibody receptor proteins, called **Fc receptors**, and to be phagocyted as well. In addition, antibodies attached to the surfaces of the bacteria act against them by facilitating the activation of complement on their surface, which, in addition to increasing opsonization, since, as we have seen, complement proteins also help to coat the bacteria, can lead to their death by forming pores in their membrane.

The third phenomenon that happens is really astounding. B lymphocytes set in motion a process of evolution and selection leading to their improvement. This process is called **somatic hypermutation** and consists of the generation of multiple daughter B cells of the originally activated one. These daughter cells possess mutations in the antibody genes and only in these. This process takes place in the lymph nodes, in a second phase of cell division, after the initial cell division that only generates cells identical to the original, as we have said. The mutations, moreover, are concentrated in the area of the antibody gene that contains the information to generate the part of the receptor (and the antibody) that binds to the bacterial or viral molecule. These mutations can cause mainly one of two effects: they can generate receptors that bind the antigen that was initially detected worse than the original receptor, or they can generate receptors that bind the antigen better than the original receptor. The cellular progeny of the original cell will therefore be varied and consist of cells that will bind less well or bind better to the antigen. Under these conditions, competition for survival takes place in the lymph node. Only those B lymphocytes that can bind the antigen strongly enough will be able to survive. Cells that cannot bind to the antigen with greater force than the initial one will die by a process of programmed cell death (i.e. consisting of a series of steps, like a computer program or an automatic washing machine). This cell death process is called **apoptosis** and is a fundamental process for the proper functioning of the immune system. We will see why later. For now, trust me and let's move on.

After a first cycle of mutation and selection, we have a new population of B cells, all of which bind better than the original to the antigen. At that time, another series of mutations occurs that will generate B cells that bind the antigen with greater or lesser strength. Again, only those that can compete successfully and bind the antigen more strongly survive. After several repetitions of these cycles of mutation and selection in the genes of the antibodies, the lymph node ends up generating a population of B lymphocytes that secrete antibodies capable of binding to the antigen with much more strength than the original, so they are much more effective.

In this way, we can see that it is not necessary to have B lymphocytes that initially bind to an antigen with great strength. It is enough for them to bind strongly enough for a B cell to be stimulated to divide. If this happens, the antibody maturation process will lead to the generation of a much more effective antibody than the original one, which will no longer be produced, and its improved version will be produced instead. In addition, the antibodies produced can be of up to five different classes, which perform different functions. Later, we will look more closely at how B cells decide which class of antibody to produce and at the defense functions performed by the different classes of antibodies.

Thus, we see that the lymph that reaches the lymph node carrying antigens from the bacteria in the source of infection, and even the bacteria themselves (or also viruses or fungi), plays a fundamental role for the generation of antibodies by the B lymphocytes. Let us now turn to the process of activation of T lymphocytes.

### 2.5.4.- Activation of T lymphocytes

As we have said, dendritic cells and macrophages that have captured bacteria or viruses in the point of infection are also transported to the lymph node. The bacteria or virus these cells have captured have been partially digested and the protein fragments generated by this partial digestion will thus serve to present the T lymphocytes with the molecular characteristics of the enemy in a safe manner. It is this function that makes it possible to call dendritic cells and macrophages by the generic name of **antigen-presenting cells**, although only dendritic cells are "professionally" and exclusively dedicated to this function, while, as we shall see, macrophages perform additional functions. Whichever cell presents the antigens, the enemy is presented to the T lymphocytes in a fragmented and dead form. T lymphocytes are the most important cells of the adaptive immune system.

We see here an important difference in the way B and T lymphocytes detect antigens. B lymphocytes detect antigens directly; they do not need other cells as intermediaries to detect them. T lymphocytes, on the other hand, do need other cells as intermediaries to detect antigens and cannot do so and be properly activated without the assistance of antigen-presenting cells. We will see later that this is related to the fact that T cells only detect antigens that have modified some of our own molecules in some way, indicating an invasion by some type of microorganism that intends to parasitize us.

Some macrophages may keep fighting at the source of infection, phagocyting bacteria, and not travel to the lymph nodes, but the dendritic cells that have phagocyted bacteria or other microorganisms perform only one fundamental mission: to travel to these nodes and present in them the antigens of the microorganisms that have phagocyted to naïve T lymphocytes, activate them and "educate" them so that they develop into **armed and effector T lymphocytes** to fight the threat in question. When we speak of armed effector cells, we mean simply that they are cells that carry out a certain defense function that produces an effect. For this reason, they are effector cells. Similarly, to carry out any mission, they need to have weapons, that is, they need to be armed. This may seem complicated to understand. What are the weapons of the effector cells? They are simply molecules produced by certain genes after the activation of these cells, and which are not produced if the cells are not activated. These molecules enable the lymphocyte to carry out a defensive function that it could not previously perform, such as killing cells infected by viruses. Activation occurs after an encounter with a dendritic cell (or also with a macrophage or a B lymphocyte) that has previously encountered an antigen.

#### 2.5.4.1.- Effector weapons

Let's go back to the antigen-presenting cells. These, once they have phagocytized the bacteria and are digesting them, begin to change the set of genes they have in operation. First, they turn off the genes that allow them to adhere to the tissues on the periphery of the organism, to the skin, for example. This causes them to detach from these tissues and be carried away by the flow of lymph that has been accumulating in the source of infection, thanks to the cytokines and chemokines that they have also produced after detecting the microorganisms. In this way, dendritic cells end up inside the lymphatic vessels and travel to the lymph nodes.

During the journey through the lymph vessels, dendritic antigen-presenting cells continue to modify the expression of numerous genes. These genes prepare them to perform their function, which is, as we have said, to present the T lymphocytes with fragments of the enemy that they have captured.

Among these genes, there are some that allow dendritic cells to increase their surface area, that is, to increase the extension of their cell membrane. In this way, dendritic cells can place on their surface a greater quantity of molecules derived from the enemy that they have detected and captured, thus increasing the probability that a lymphocyte can detect them. The increase in the surface of the membrane causes it to fold and form structures like the dendrites found in neurons. This is the reason why these cells are called dendritic cells and not because they have anything to do with neurons. This ability to increase their surface area is one of the properties that makes dendritic cells specialized for the function of presenting antigens. Although macrophages also present antigens, as they do not increase much their surface area, they cannot achieve the level of presentation efficiency that dendritic cells do.

We have already stressed that, if the cell is to perform one or more functions, it needs molecular tools to do so. In this case, the mission of antigen-presenting cells is to present on their surface small fragments derived from enemy proteins, of only and average of seven to eight amino acids in length, although some may be longer. These small protein fragments are called **peptides**. To generate these peptides, the cell needs a complex enzymatic machinery that partially digests the captured enemy proteins. In fact, this machinery also serves to degrade the cell's own proteins that have been aging (denaturing) and need to be recycled. This complex machinery is called the **proteasome**, and it is made up of about twenty-eight proteins. The proteins that are components of more complex systems are called **subunits**, so the proteasome is composed of at least twenty-eight protein subunits. As a point of comparison to estimate the complexity of this machinery, let us consider that the hemoglobin in our blood is formed by only four subunits, although the ribosomes, the cellular organelles in charge of protein synthesis, are formed by the assembly of eighty-two subunits.

Certain subsets of dendritic cells and some macrophages are very efficient in capturing antigens from the extracellular medium and digesting them. Capture occurs by phagocytosis or macropinocytosis. The ingested antigens are either transported to the cytosol for degradation in the proteasome, or digested in the phagocytic vesicle itself, called the **phagosome**, which then fuses with another digestive vesicle called the **lysosome**. As we know, the suffix *some* means 'body'. Thus, phagosomes are bodies derived from phagocytosis and lysosomes are bodies that lyse, capable of lysing and digesting, because they contain numerous digestive enzymes. Both the phagosome and the lysosome are vesicles, that is, small spheres located inside the cells, formed by a cell membrane that separates their contents from the rest of the cytoplasm. In the case of lysosomes this is essential to prevent the digestive enzymes they contain from digesting the cell itself. The fusion of a phagosome with a lysosome brings the digestive enzymes of the lysosome into contact with the antigenic particles captured in the phagosome and digests them. Once these particles are digested, the resulting peptides will be used to be presented to T lymphocytes.

Antigen-presenting cells activated by the detection of an enemy also increase the expression of the genes that produce some of the proteasome subunits, especially those that increase their efficiency in digesting the proteins derived from the enemies and in producing, from them, the peptides to be presented on the membrane to the T lymphocytes. Likewise, activated antigen presenting cells antigens increase the expression of the genes to produce the proteins that must capture the peptides produced by the proteasome and place them on the membrane, so that some T lymphocyte can detect them. These genes are fundamental for the function of the immune system and are the ones that constitute the so-called **major histocompatibility complex**.

Histocompatibility means compatibility among tissues and organs. Later, we will explain where this name comes from, as well as how the proteasome works, but now this would only lead to us being too distracted from the process of activating the T lymphocytes. This process, like any process of cell activation, depends on the lymphocytes detecting an external molecule with some receptor on its cell membrane. We have seen before that B cells were activated in this way, and practically all cell activation processes occur in the same way: a molecule is detected by a receptor, which causes changes in the functioning of genes or the assembly of some molecular complex from already manufactured components that execute an action. Some genes begin to be expressed, while others are silenced. This changes the capabilities of the cells, which now do something they couldn't do before, or stop being able to do something they did before.

In the case of T lymphocytes, some of them, when activated, acquire a "license" to kill our own cells. Interestingly, killing our cells is the most effective defense mechanism when they have been subverted by enemies who put them at their service. These enemies are, in general, viruses, which use the cellular machinery to reproduce themselves and generate hundreds of new viral particles that could infect as many cells. Remember that bacteria grow in geometric proportion, doubling in size every few minutes. Viruses also reproduce in geometric proportion, but their reproduction is even faster than that of bacteria, because from one virus not only two new viruses are generated, as in the case of bacteria, but even hundreds. This is the reason why the immune system needs to be truly expeditious in the case of viral infections. We will explore this issue further below.

Let's go back to the activation of T lymphocytes by antigen-presenting cells. As we have said, these possess specialized proteins that bind the protein fragments derived from the partial digestion of phagocytized bacteria, or also from viruses that have been able to be internalized in the process of micropinocytosis, and place them on their membrane. These proteins are produced by the genes of the **major histocompatibility complex**, known as **MHC**. Let's focus on how these proteins work.

There are two main classes of these genes and proteins, which are called **class 1 MHC** (MHC-1) and **class 2 MHC** (MHC-2). They fulfil different missions. MHC-1 genes generate proteins that place on the cell surface peptides generally derived from the degradation of **proteins that have been synthesized in the cytoplasm by the cell itself**. As a result, MHC-1 proteins place on the cell membrane peptides from viral proteins, since viruses, in order to reproduce, take control of the cellular protein synthesis machinery and thus it is the cell itself that produces the proteins that make up the virus. In this way, a cell infected by a virus will show on its surface, attached to MHC-1 molecules, peptides foreign to itself, coming from the virus proteins that have been produced in the first place and partially digested later.

MHC-2 molecules also present peptides on the membrane, but these **come from proteins that have been ingested by the cell**, either by phagocytosis, macropinocytosis or endocytosis (a process in which proteins must first bind to a specific receptor before they are ingested). Obviously, this means that dendritic cells, and other cells as well, possess fascinating mechanisms that make it possible to distinguish the origin of a protein, especially to distinguish whether it has been produced by the cell or ingested from outside.

T lymphocytes waiting to be activated, **called naïve T lymphocytes**, are found, in general, in the lymph nodes, inside the so-called **T-cell zones**, which are obviously areas of the nodes where T lymphocytes accumulate, attracted there by some chemokines. Just as there are billions of different B lymphocytes, each with a particular receptor (also called **BCR**, B-cell receptor), there are billions of different T lymphocytes, each with a given T receptor (also called **TCR**, T-cell receptor). The important difference between them is that while B-cell receptors can detect virtually any type of substance, T-cell receptors are **restricted** to detecting only peptides bound to MHC molecules. This restriction is also very strict, since the T lymphocytes that a person has generated in his lifetime will detect peptides attached to his own MHC molecules, but, in general, they will not detect those same peptides attached to other people's MHC molecules. Thus, the immune system can perform one of its most important tasks: to answer the question what is self and what is foreign? Self must be tolerated and protected; foreign must be attacked and eliminated. MHC molecules are the main, but not the only, responsible for making this molecular feat possible.

Millions and millions of T-lymphocyte precursors are generated in the bone marrow (each one with a specific receptor that we will later see how it is produced) and matured and "selected" in the thymus (we will later see how this very important selection process is accomplished). Once mature, T lymphocytes abandon the thymus via the blood and penetrate to the lymph nodes through the blood vessels, in a process very similar to that used by monocytes and macrophages to leave the blood and reach the sources of infection. Once inside a lymph node, naïve T lymphocytes gather in the so-called T zones of the lymph nodes, which are located throughout the body's anatomy, although most are found surrounding the intestine. To the lymph nodes also continuously arrive dendritic cells carried there by the lymph. In this case, these cells come through the **afferent** (ingoing) lymphatic vessels. Once in the lymph node, these dendritic cells penetrate the T-lymphocyte zones. Each dendritic cell carries on its surface thousands of MHC molecules of both classes attached to different peptides, which come from different fragments of the many proteins expressed by bacteria or viruses encountered at the source of infection. Therefore, each dendritic cell displays on its surface a rich and diverse collection of peptides attached to MHC-1 or MHC-2 molecules. Many of these peptides, however, come from the cell's self-proteins that have been degraded and cut into peptides, or from the body's self-proteins, captured by the dendritic cell and degraded also into peptides. Thus, a dendritic cell and other cells that present antigens to T cells, among which, as we will see, are also macrophages and B cells, display on their surface an extensive collection of MHC molecules linked to both self and foreign peptides. The presence of the latter will lead to the activation of naïve T lymphocytes.

At this point in the story, the naïve T lymphocytes have not yet found the love of their life, those dendritic cells that will activate them and thus change and give meaning to their existence forever. The naïve lymphocytes accumulated in the T zone of the lymph nodes will come into contact with the surface of the arriving dendritic cells and will explore this surface, sliding over it, in search of a peptide attached to an MHC-1 or MHC-2 that is capable of binding strongly to its receptor. This strong binding (with high affinity, as it is said in scientific language) is important, since a weak binding, if it occurs, does not lead to the activation of T lymphocytes. Later, we will explain why the binding affinity between the T-lymphocyte receptor and peptides bound to MHC molecules is so important for their correct activation. For now, let's clarify that most T lymphocytes will not find that peptide attached to an MHC that will activate them. They will continue being naïve and, after a while, they will leave the lymph node through the **efferent** (outgoing) lymphatic vessels and will return to the blood through the thoracic duct.

Once again in the blood, they will re-enter another lymph node in search of the dendritic cell that could activate them. Most naïve T lymphocytes, therefore, will wander around the body's lymph nodes all their lives without finding an antigen presented by a dendritic cell that will activate them. Many will die as itinerant naïve lymphocytes, since they will never find the dendritic cell with a peptide that could have activated them.

To complicate things a bit further, it turns out that there is not just one type of T lymphocytes, but two. The main difference between them is that the first type of T lymphocyte has receptors that detect exclusively peptides attached to MHC-1 molecules. These are the so-called **CD8 T lymphocytes**. The other type of T lymphocytes has receptors that exclusively detect peptides bound to MHC-2 molecules. These are the so-called **CD4 T lymphocytes**. Let us remember this, because it is important: the relationship is **MHC-1/CD8**, **MHC-2/CD4**. Later, we will explain why the main types of lymphocytes are called CD4 and CD8, but, for the moment, let's just remember that each of these T-lymphocyte classes binds to an MHC molecule also of a given type.

As we have said, dendritic cells express on their surface both peptides bound to MHC-1 and peptides bound to MHC-2, whose origin is different: MHC-1 binds peptides produced mainly by the cells themselves and MHC-2 binds peptides captured from the outside. A dendritic cell will therefore be able to activate both types of lymphocytes; to do this it must find one that has a receptor that strongly binds to one of the peptides it has attached to an MHC-1 molecule or an MHC-2 molecule.

However, let's consider that there are thousands and thousands of peptides from microorganisms and that there are billions of T lymphocytes, each of them with a receptor capable of detecting a different peptide attached to an MHC molecule. If a dendritic cell has captured, for example, a bacterium and presents peptides of the bacterial proteins in its MHC-2 molecules, the dendritic cell will present hundreds or thousands of peptides coming from the bacterium. It is, therefore, almost certain that at least one of the T lymphocytes will recognize, that is, will be able to bind through its receptor, to one of the bacterial peptides presented by the dendritic cell, and the T lymphocyte will be able to be activated by it. The process is like playing bingo, like a lottery that one or another always wins: Is there any lymphocyte that has a receptor for any of the peptides presented? Since there are a lot of peptides and a lot of lymphocytes, one or another T lymphocyte always wins this lottery, even more than one of the T lymphocytes may win.

Later we will see in more detail what missions the different types of activated T lymphocytes will accomplish, but let's now look a little more closely at the process of their activation. The binding of the T-lymphocyte receptor to an MHC molecule with a peptide triggers a series of biochemical mechanisms that send the signal to the interior of the lymphocyte. These mechanisms finally result in modifications in the functioning of the genes of this lymphocyte, as we have already said is the case in all cell activation processes. However, this signal from the lymphocyte receptor (for which transmission to the cell nucleus requires the participation of CD4 or CD8 molecules, depending on the type of lymphocyte) is not the only one necessary for the lymphocyte to be activated. The lymphocyte must receive two more signals from the dendritic cell to be fully activated. Let's see what these two additional signals are and why they are necessary.

The second signal required for the T lymphocyte to be activated is provided by a pair of molecules also present on the surface of dendritic cells activated by their encounter with a microorganism. These molecules bind to a different receptor present on the T-lymphocyte membrane. As in the case of the binding of MHC molecules to T-lymphocyte receptors, this binding will send a signal to the interior of the cell that will help set in motion the genes necessary for the T lymphocytes, once activated, to carry out their mission. These two molecules present on the surface of dendritic cells are called **co-stimulatory molecules** and are referred to by the specific names of **B7-1** and **B7-2**, also known as **CD80** and **CD86**. The genes that produce these molecules are set in motion after activation of the dendritic cell that has captured an enemy and during the cell's journey from the periphery to the lymph nodes through the lymph vessels. In this way, when they reach the lymph nodes the dendritic cells are prepared to correctly activate the T lymphocytes that can first be stimulated by binding to an MHC with its peptide, which the dendritic cell will also present on its surface. The B7-

1 and B7-2 receptor is expressed on the surface of the T lymphocytes and is called the **CD28 co-receptor**.

The need for this second signal so that T lymphocytes are fully activated is what prevents T lymphocytes that could recognize self-peptides attached to MHC molecules from being activated by any cell other than the dendritic cells that express these complexes. These cells from other organs and tissues will not express on their surface either CD80 or CD86, so they will not be able to send co-stimulatory signals to those lymphocytes that could detect the self-peptides presented by MHC molecules. Normally, these would be, in addition, self-peptides that should not stimulate any attack by the immune system. Therefore, in the case that a T lymphocyte detects those peptides with enough strength, but in the absence of the simultaneous expression of co-stimulatory molecules, that is, when it receives a signal through its antigen receptor without receiving at the same time the co-stimulatory signal, the T lymphocyte not only will not be activated, but it will enter in a 'sleepy' state in which it cannot be activated anymore. This state is called **anergy**, and it is an important mechanism of tolerance (inhibition of the attack) to self-antigens. Also, on other occasions, the T lymphocyte that identifies an antigen in the absence of co-stimulation may induce its own suicide by apoptosis. This process is called **clonal elimination**, because it eliminates those T cells that could give rise to clones of lymphocytes capable of attacking our own body. We will see later that this is a fundamental process during the development of T lymphocytes. In conclusion, the correct activation of T lymphocytes always takes place in a cellular context of inflammation, that is, thanks to dendritic cells activated by signals emitted by Toll-like receptors that have detected molecules of the microorganisms, and which have been carried by the lymph to the lymph nodes. Only these dendritic cells or activated macrophages express CD28 and can correctly stimulate T cells.

The letters CD that always precede a number indicating one or another molecule of the immune system, come from **cluster of differentiation**. This CD group is made up of molecules that, over years of research, have been identified as being able to help distinguish, or differentiate, some immune system cells from others. Thus, for example, a cell that expresses CD28 is a T cell, but a cell that expresses CD80 will be a dendritic cell or other similar cell capable of activating T lymphocytes. The expression of one or another CD molecule allows the identification of the different classes of cells in the immune system. The CD group consists of more than 370 proteins. Likewise, over the years, the functions that each of the CD molecules perform in the different cells expressing them have been discovered. However, at this point, this is just an unnecessary curiosity to understand the functioning of the immune system. So, let's move on to the third signal, the third stimulus necessary for the correct activation of T cells.

This third signal is supplied by dendritic cells through the production of proteins which, in this case, are secreted to the external environment and which receive the generic name of **cytokines**. This is the only signal that does not depend, therefore, on the direct interaction between the two cells, the dendritic cell and the naïve T cell, although the cytokines are released in a directed way into the space between the two interacting cells. The first two signals are produced when these two cells literally touch, thus allowing MHC and T-cell receptors, and co-stimulatory molecules and the CD28 coreceptor, to bind and T lymphocytes to thus receive these first two signals. Once this binding is produced, the dendritic cell secretes some cytokines, most of them in the small space between the two contacting cells.

We've encountered chemokines before. The word "cytokines" also ends in the suffix *kyne*, which, let's remember, comes from ancient Greek and means 'movement'. On the other hand, the prefix *cyto* also comes from the Greek and means 'cell'. Therefore, cytokines are substances that make cells "move", but in this case the movement is not in space; it is not a transfer out of the blood vessels or a patrolling of the tissues, but a change in the cells' capabilities. As we already know, the changes are going to be due to specific modifications in the functioning of certain genes that are going to allow the cells to do now things that they could not do before. Some, but not all, cytokines are also called by the generic name of **interleukins**. We will see that the specific name of many of them begins with the letters IL (interleukin), followed by a number for each cytokine: IL-1, IL-2, etc.

Cytokines can be classified into four broad families composed of structurally and functionally related cytokines. Each cytokine has a specific receptor for it on the membrane of the cells that can detect it. Cytokine receptors can also be classified into related families and, according to the cytokines that stimulate them, each receptor will be able to send a different signal to the cell nucleus, which will turn certain genes on or off. Therefore, the action of the cytokines will help to transform the activated cells into cells different from the non-activated naïve cells, since one cell differs from another exclusively by the set of genes it expresses. The action of the cytokines thus joins the action of the other two signals and manages to modify the level of functioning of an established set of genes that will enable the T lymphocyte to be activated and differentiated into a cell that will carry out a specific mission.

It is worth stopping for a moment a little bit longer at this third signal provided by the cytokines, necessary for the activation of the T lymphocytes. The first two signals, the one provided by the MHC:peptide complex and the one provided by the co-stimulatory molecules, are transmitted to the interior of the T lymphocytes by identical receptors and co-receptors shared by all of them. It does not matter that the MHC molecule and the bound peptide might be different and activate lymphocytes with particular receptors for them; the signal sent by these receptors to the interior of each T cell is virtually the same for all T lymphocytes that detect an MHC molecule bound to a peptide. Similarly, the signal sent by the CD28 receptor when it binds to CD80 (B7-1) or CD86 (B7-2) is the same for all T lymphocytes. The third signal, however, that provided by cytokines, is different because it depends on the specific cytokines produced by the dendritic cell that communicates with the T cell to activate it. Each of these cytokines activate a different receptor, as we have said, which will cause the activation of different genes and will lead to the T lymphocyte being activated and differentiated, that is, to acquire the molecular tools necessary to carry out the function that the dendritic cell commands, tools that will depend on the expression of the genes that the three combined signals set in motion. However, although the three signals are necessary, the signal that causes the T lymphocyte to differentiate and perform a specific function is, above all, the signal sent by the cytokines.

All of the above means that in the process of activating the T lymphocytes there are subprocesses common to all of them, initiated by the signals received through the T-lymphocyte receptor and the CD28 co-receptor, and there are specific signals received through the cytokine receptors. The three signals trigger reactions in the cytoplasm of the naïve T lymphocyte that act in an integrated and coordinated manner in the cell nucleus, where the appropriate genes are set in motion. This way, the T lymphocyte is activated, reproduces, generates a clone of thousands and thousands of identical cells, in a process called, as we have seen, **clonal expansion**, and the T lymphocytes thus activated and generated acquire the correct set of functionalities to fight the enemy in question.

And how does it know what the nature of the enemy is? Let us remember that this information has been received first by the sentinel cell, that is, the dendritic cell, through the activation of its Toll-like receptors. It is this cell that has been stimulated by the detection of the foreign microorganism molecules that it has found in the source of infection. These molecules, depending on their capacity to activate one or another Toll-like receptors and other receptors of innate immunity, communicate to the dendritic cell information about the nature of the enemy, and it tells this cell if it is dealing with a bacterium, a virus, a fungus, etc. Then, the dendritic cells is activated in a manner in accordance with that information. This activation leads to the dendritic cell secreting one or another cytokine when it encounters a T cell capable of being activated, which will depend on whether or not the receptor of this cell can bind with enough strength the MHC:peptide complexes presented by the dendritic cell. In this way, the cytokines are the way in which the dendritic cell communicates to a T cell recognizing the MHC:peptide complexes that the dendritic cell presents to it the information about the type of enemy it has encountered.

The information about the nature of the enemy that the dendritic cell or macrophage have identified, however, is also communicated in part by the type of MHC molecule used by the dendritic cell to present foreign peptides to the naïve T cell that must be activated. If the dendritic cell has been infected by a virus or has captured virus particles, it will present peptides derived from viral proteins on the MHC-1 molecules. In this case, only naïve T lymphocytes with receptors capable of recognizing MHC-1 molecules loaded with peptides can be activated. These T cells will only be those that express the CD8 co-receptor on their membrane. Let us remember that this co-receptor participates in the molecular signal that is sent by the T-cell receptor, but it is not an independent co-receptor, as CD28 is. CD28 has its own activating molecule (its own **ligand**, as it is said in scientific language, since a ligand binds to a receptor), but CD8 does not possess an independent ligand, as neither does the CD4 co-receptor. In fact, the CD8 ligand is a lateral zone of the MHC-1 molecule, and the CD4 ligand is a lateral zone of the MHC-2 molecule. These lateral regions of MHC-1 and MHC-2 molecules are not involved in the binding of the peptide to them. Thus, the T-cell receptor and the CD8 and CD4 molecules trap the MHC-1 or MHC-2 molecules, respectively, as if they were a clamp. Thus, the T cell and the dendritic cell are joined. Proteins that form "molecular clamps" appear with some frequency in the interaction between the molecules of the immune system. Remember that integrins **(section 2.5)** also form a kind of clamp when they interact with adhesion molecules. As we know, adhesion molecules and integrins are also needed to maintain the intercellular connection between dendritic cells and T cells during the process of T cell activation.

Let's continue with the cytokines produced by the dendritic cell and secreted towards the T cell, which detects them. The dendritic cell that has caught a virus "knows" that it must activate CD8 T cells. The dendritic cell "knows" this because it has detected through the Toll-like receptors viral molecules that have activated it in a certain way, a way that sets off particular genes to produce the cytokines that the naïve CD8 T cell needs to be activated and become a killer cell. Later we will see **(section 2.7)** how these interesting killer cells defend us from a deadly threat by killing our own cells, something that is very paradoxical, but absolutely necessary to defend ourselves from viruses, since, after hundreds of millions of years, evolution has not invented or discovered a less dangerous and costly mechanism.

A similar process occurs when the dendritic cell has detected other types of microorganisms, for example bacteria, and must activate CD4 T cells. This is when things get complicated, but let's try to keep them at an affordable level and go step by step making the processes and reasons for them very clear.

The reason why things are more complicated in this case is because, first, there are several types of bacteria that, from the point of view of defense mechanisms against them, we can basically divide into two main kinds: those that live outside the cells (extracellular) and those that live inside them (intracellular). These two types of bacteria need different molecular and cellular tools to fight them effectively. These molecular tools are put to work by activating naïve CD4 T cells and differentiating them into two main types of armed effector CD4 T cells: the so-called $T_H1$ cells and the so-called $T_{FH}$ cells. Let us immediately clarify that these two are not the only types of CD4 T cells that exist. There are several more, and we will meet them later. Let's focus, however, on $T_H1$ and $T_{FH}$ cells now. $T_H1$ cells orchestrate and direct the fight against intracellular bacteria, but they can also participate in the fight against viruses, helping CD8 T cells and B cells to generate neutralizing antibodies against them. $T_{FH}$ cells organize and direct the fight against extracellular bacteria and against viruses, helping B cells to generate the right kind of antibodies that neutralize them and facilitate their phagocytosis and elimination.

The ability to help other cells is what gives $T_H1$ and $T_{FH}$ cells their names, since the letter "H" that appears on them is the initial of the word "helper". The number 1 in $T_H1$ indicates that these cells are the first type of helper cells. As we have already said, there are a few more. The letter "F" in $T_{FH}$ indicates that this type of cell resides in a specific zone in the lymph node: the **lymphoid follicle**. Later **(section 6.1),** we will explain in some detail what the follicle is and how immune cells are organized in it.

#### 2.5.5.- The spleen and its function

I think it is appropriate, now that we have learned the basics about lymphocyte activation, to take a short break to address an issue that may have troubled some readers. Although most infections are localized in tissues, what happens if some bacteria get into the blood and spread through it? It is not the case here that the bacteria, because the infection could not be controlled, invade the blood and cause septic shock. It's about how to control the few bacteria that do get into the blood when we are injured so that they don't grow in the blood and invade the whole body, causing widespread infection. Those few bacteria must be equally eliminated. How is this achieved?

To begin with, it is necessary to clarify that in the daily lives of people who are careful with their oral hygiene, bacteria enter the bloodstream every time we brush our teeth, whether our gums bleed visibly or not. Since brushing your teeth is not a deadly activity, but on the contrary a healthy action, it is obvious that the bacteria that enter the blood are neutralized and eliminated without our knowledge. The organ that plays a key role in this, especially during childhood, is called **the spleen**.

The spleen is an organ located on the left side of the abdomen, towards the front, and at the level of the liver, above the intestine. Its structure is partially similar to that of a large lymph node, but instead of receiving, like these, antigens from the lymph, the spleen receives them directly from the blood, filtering it and bringing its contents into contact with phagocytic cells residing in it. These phagocytic cells perform the function of eliminating aged or abnormal cells, mainly red blood cells, and eliminate also microorganisms that may have penetrated the blood. In the latter case, these phagocytes become antigen-presenting cells and activate T lymphocytes present in the T zones of the so-called white pulp of the spleen, which also has zones of B lymphocytes that can also be activated if they detect an antigen specific to them. The white pulp is located around the small arteries that form the network capable of filtering the blood together with the veins that collect the blood supply.

The production of antibodies by spleen B cells is an important function to neutralize and opsonize microorganisms that have been able to penetrate the blood and that have not been able to be completely eliminated by passing through the so-called red pulp, which contains most of the phagocytes of the spleen. In fact, some bacteria are coated with a capsule made up of carbohydrates (polysaccharides). For this reason, these bacteria cannot be eliminated unless they are coated with antibodies. In addition to normal B lymphocytes (also called B2 lymphocytes), there are certain B lymphocytes, called **B1 lymphocytes**, which are responsible for the production of antibodies against this type of bacteria. Once coated with them, these bacteria are effectively killed by phagocytes in the spleen.

Although they produce antibodies, B1 lymphocytes are considered cells of the innate immune system. This is because the antibodies produced by these cells are usually of the IgM class **(section 2.9)** and bind to repeating molecular patterns present on the surface of the bacteria. Furthermore, for the generation of these antibodies, B1 lymphocytes do not require the assistance of T helper lymphocytes. B1 lymphocytes also cannot become memory B cells (cells that remember the type of microorganism they have encountered before and are activated much more rapidly than naïve lymphocytes), which, as we will see **(section 6.1),** is one of the most important features of adaptive immunity. B1 lymphocytes are not only found in the spleen, and they are abundant in the pleural (lungs) and peritoneal (intestine) spaces.

The function of the spleen is more important during childhood, since the organism has not yet found all the microorganisms that are typical of the environment in which it lives and has not yet generated memory cells to defend itself from them. Once the memory cells have been generated, i.e. when the body has been naturally vaccinated against bacteria and other organisms that have entered the blood, the function of the spleen is no longer as important in controlling them. For this reason, this organ can be removed if necessary, such as if it ruptures as a result of a traumatic accident. In addition, once coated with antibodies, which can be produced by activated memory B lymphocytes in any lymph node, macrophages present in the liver are also effective in removing bacteria from the blood.

For the above reasons, in the event of removal of the spleen, physicians recommend the administration of vaccines against the encapsulated bacteria that cause the most common diseases. Preventive antibiotic prophylaxis is also recommended before any procedures that facilitate the entry of bacteria into the blood, such as dental or surgical procedures.

## 2.6.- $T_H1$ Lymphocytes and macrophage activation

Let's now return to the $T_H1$ lymphocytes. What are their functions? Simplifying the complexity of the world in which these T lymphocytes live, we can say that they perform two main functions: the first is to help macrophages eliminate the bacteria they have phagocytized; the second is to help CD8 T cells overcome their reticence and finally have the molecular arrests to start killing infected cells of our body, because our lives depend on it.

To understand the importance of the first function, we must return to the source of infection. Let us assume that the bacteria that form it in this case are the type that prefer to live inside the cells. This bacterial lifestyle may have certain advantages, because in this way the bacteria avoid the action of antibodies and complement, molecules that are always found on the exterior of the cells and cannot reach the cytoplasm of any cell in the body. However, the interior of the phagocytes is not a very hospitable place. Phagocytes have an arsenal of chemical weapons inside them to digest bacteria. Nevertheless, some species of bacteria have "learned" throughout their evolutionary history to bypass these weapons and thus take advantage of the abundant nutrients in the cytoplasm of the cell that has phagocytized them but cannot digest them.

Throughout evolution, two defense mechanisms have been developed against bacteria that resist digestion. The first of these has been developed by neutrophils, cells that are the first to reach the point of infection in response to the cytokines sent by the sentinel cells. Neutrophils, which are very numerous, are very powerful phagocytes. In fact, they are so powerful that phagocytosis and digestion of the phagocytic bacteria often leads to their own death. Dead neutrophils accumulate in the wound in the form of pus, which, if not released to the outside, is then cleaned up and eliminated by the other most important phagocytes of the defenses: the macrophages. The short life of neutrophils prevents bacteria from developing inside them. This is an advantage, but it prevents these phagocytes from giving the alarm to other cells about the enemy they have encountered. Neutrophils are kamikaze fighters and, as such, cannot return to the barracks to inform their superiors of the outcome of the battle and the nature of the threat. This mission, on the other hand, is carried out by dendritic cells and macrophages, cells that do not die after the phagocytosis of bacteria; cells that can travel to the lymph nodes to give the alarm to the T cells and that, for that reason, have been the target of some bacteria as hosts that can be parasitized to live inside them.

These bacteria belong to the class of so-called **mycobacteria**, which cause diseases as serious as tuberculosis or leprosy, diseases which, although rare in developed countries, persist in less favored countries, where they cause significant mortality. In fact, data published by the World Health Organization in 2016 place tuberculosis as the tenth cause of death in the world, although it is the seventh cause of death in developing countries. Therefore, the control of mycobacteria has been, and continues to be, an important factor in the evolution of the immune system towards the development of effective mechanisms to eradicate them. These mechanisms are often ineffective and mycobacterial infections cannot be controlled, leading to death, as is always the case when an infection cannot be stopped.

However, macrophages have developed over the course of evolution very powerful and effective mechanisms to eliminate mycobacteria that have parasitized them, but these cells cannot set them in motion without the "permission" of the activated $T_H1$ cells. This is a rather curious fact, or at least it seems so to me. It turns out that macrophages that phagocytize some species of bacteria cannot kill them inside unless other cells have been warned and allow them to set in motion the mechanisms that will kill them. The obvious question is: Why?

The answer can be understood if we consider that the activation of these mechanisms will cause, as we have said, a collateral damage that can be serious. To understand this, let us suppose that, in a war, a dangerous enemy invading a city or a region, which could end up invading the whole country, could be eliminated by using a very powerful bomb, let us suppose a nuclear bomb. This bomb would kill the enemy, but it would also finish with the lives of tens of thousands of our compatriots. This is a decision that cannot be taken lightly, nor should it be taken by a single person.

The situation in which macrophages that have phagocyted mycobacteria capable of surviving the regular mechanisms of their elimination that these cells set in motion is like the previous one. In order to kill them, macrophages need to activate more powerful weapons, but these will not only kill the bacteria, but also damage the surrounding cells and tissues. This decision should not be made by macrophages alone, as macrophages could make mistakes and activate these mechanisms in an inappropriate or unnecessary way, which could end up favoring the enemy they are trying to eliminate.

For this reason, we need the action of $T_H1$ lymphocytes. The naïve CD4 T lymphocyte that generates a clone - that is, thousands of cells identical to the original - of activated $T_H1$ cells has been stimulated in the lymph node by a dendritic cell or by a macrophage that has gotten there and presented to it a bacterial peptide attached to an MHC-2 molecule. The dendritic cell or macrophage has, in turn, been activated by the microorganism that was encountered at the source of infection. The molecules of this microorganism, through the Toll-like receptors, set in motion the genes necessary to produce the cytokines that will induce the differentiation of the naïve CD4 T cell to a $T_H1$ cell. One of the most important cytokines for this purpose is **interleukin 12 (IL-12)**. The peptide presented by MHC-2 molecules comes from a bacterium at the site of infection, where macrophages and neutrophils continue to fight, with the help of complement, by phagocyting the bacteria. Once activated, the $T_H1$ cells set in motion new genes and turn off those that produce the proteins that kept it anchored to the lymph node. Now these cells no longer have any grips and can leave the node through an efferent (outgoing) lymph vessel. After traveling through the lymph vessels, the cells join the bloodstream through the thoracic duct. Once in the blood, the $T_H1$ lymphocytes will circulate through the blood vessels and it will only be a matter of time before they end up passing through one close to the source of infection. Let us remember that the endothelial cells of these blood vessels have been activated by the cytokines secreted by the phagocytes and dendritic cells that had encountered the bacteria. This activation makes them sticky to the phagocytes, but they are equally sticky to the activated $T_H1$ lymphocytes (although not to the naïve CD4 T lymphocytes). They can adhere to the surface of endothelial cells near the site of infection and since the joining between these cells is also relaxed, they can pass through the endothelium between two endothelial cells and exit into the surrounding tissue in search of the site of infection. Similarly, the activated $T_H1$ cells can now "smell" the chemokines secreted by the phagocytes that continue to fight at the site of infection and, guided by that "smell", navigate through the tissue until they reach the point of infection.

The site of infection is where macrophages that have phagocyted mycobacteria, but cannot eliminate them, are found. Mycobacteria can bypass the molecular mechanisms macrophages put in place for digesting microorganisms. However, some mycobacteria inside the macrophages have been partially degraded, so that the infected macrophages present on their surface bacterial peptides linked to MHC-2 molecules. Some of these peptides are identical to those that have activated $T_H1$ cells in the lymph node and these cells will be able to detect them. When this happens, $T_H1$ cells "know" that they are in contact with an infected macrophage that they must help.

The help of $T_H1$ cells to infected macrophages results in two actions. The first is the interaction of a molecule on the $T_H1$ membrane with a macrophage membrane receptor. This receptor is one of the most important for the functioning of the immune system and is called **CD40**. The ligand of this receptor, not surprisingly, is called **CD40L** ("L" is obviously the initial of the word "ligand"). The CD40L molecule is expressed by activated $T_H1$ cells. With it, these lymphocytes will stimulate the CD40 receptor of macrophages. As we have already said, whenever a receptor is stimulated a series of molecular events occur in the cytoplasm of the receptor cell. Some converge in the nucleus, where certain genes are activated or silenced. Others can activate resting molecules in the cytoplasm, which are then activated by receiving the signal from the receptor, usually leading to chemical modifications in the target molecules. This is what happens in this case. The interaction between the CD40L molecules expressed by $T_H1$ cells and the CD40 receptor molecules, expressed by the macrophage, causes the activation of molecular mechanisms in the macrophage that will allow this cell to perform functions it could not perform before. Among these functions is the activation of very powerful bacterial attack mechanisms.

The second action carried out by the $T_H1$ cell is the generation and secretion of certain cytokines. One of the most important is the so-called **interferon-gamma (IFN-γ)**. Macrophages that are infected and activated by phagocytic bacteria always have receptor molecules for IFN-γ on their surface, but if these receptors do not detect anything, if no $T_H1$ cell helps them, the macrophage does not do anything either. It continues to try to control the bacteria inside in the least painful way possible for the body, a way that, however, in some cases will not succeed in eliminating the mycobacteria completely.

The physical interaction between the $T_H1$ cell and the macrophage via CD40-CD40L interaction, the secretion of IFN-$\gamma$ by the $T_H1$ cell and its detection by IFN-$\gamma$ receptors present on the surface of the macrophages sends to these cells very important information, at the same time that it transmits a command to them. The information macrophages receive when they detect the cytokines secreted by $T_H1$ cells is that they now "know" that at least one naïve T cell has been activated in the lymph node by a dendritic cell or other macrophage that has traveled there and presented to that naïve T lymphocyte a peptide identical to the one that macrophages in the source of infection now present to the activated $T_H1$ cells. This indicates to the macrophage that another cell, in this case, a dendritic cell or another activated macrophage, has also detected the same enemy, has been activated, has migrated to a lymph node and has been able to activate a naïve CD4 T cell and make it mature towards the $T_H1$ type and divide, generating many identical cells that now come to the source of infection in aid of the macrophages.

In addition to this information, the CD40L and the IFN-$\gamma$ transmit a command to the macrophages: the order to activate themselves to their maximum potential and to destroy the enemy inside them. Macrophages do this in two ways. One is by stimulating the digestive enzymes to reach the bacteria and digest them; the other is by generating very powerful **oxidizing substances**, such as superoxide anion and hydrogen peroxide, which oxidize the bacteria and kill them. Remember that the process of generating these substances is called a **respiratory burst (section 2.2)**. These compounds, however, are also toxic to our own cells and tissues and this is why, before using them, macrophages must ensure that it is the correct action that needs to be carried out, which only happens if indicated by $T_H1$ cells activated by peptides derived from the same microorganism.

However, not all macrophages parasitized by bacteria will be able to kill them even with the help of $T_H1$ cells. The reason is that many of these macrophages may have been damaged by the bacteria, or the bacteria may have produced molecules that prevent or slow down the production mechanisms of the toxic substances that would kill them. In this case, the cytokines produced by $T_H1$ cells will be useless, since they will not be able to act adequately on the macrophages they are supposed to activate. Fortunately, there is another mechanism by which the bacteria phagocyted by macrophages can be eliminated, even if the macrophages cannot activate the processes for their digestion. This mechanism involves the suicide of the infected macrophage by the process of programmed cell death, called apoptosis, which we mentioned a few pages ago.

It is worth stopping briefly to look at what the process of cell apoptosis or suicide involves and what benefits it provides for the survival of the whole organism. To clarify, let's say that cell suicide can occur for two reasons. The first is that the cell is not capable of obtaining the material and energy resources to continue living. In this case, it is the cellular mitochondria, the organelles responsible precisely for the generation of chemical energy necessary for the living processes, which emit to the cellular cytoplasm some molecules that trigger the process of suicide. The second reason is that the cell receives a suicide order from another cell interacting with it by sending molecules that bind to the so-called **death receptors**, which trigger the suicide process. This "superior order" suicide and the existence of these "death receptors" on the cell membrane, always ready to receive the suicide order if necessary, is one of the most amazing processes in the functioning of the immune system and places death in an intimate relationship with life. One of these death receptors is the one called Fas, which is activated by a membrane molecule, its ligand, called Fas ligand, or FasL. $T_H1$ lymphocytes express FasL on their membrane and can thus induce death by apoptosis of macrophages unable to kill by means of the respiratory burst the bacteria they have phagocytosed, which express the Fas receptor.

Death by the process of apoptosis is clean and orderly, so it has minimal impact on the rest of the living cells in the body. The cell enters a process of self-digestion in which DNA and proteins are broken down into small pieces and the cell dies, but its remains are kept inside the cell membrane, without the membrane breaking down. This generates a kind of membranous sarcophagus that keeps all the cell contents inside. These contents include the bacteria that the dead macrophage could not digest, which prevents them from spreading throughout the body and allows other healthy, younger macrophages to now phagocytize the dead macrophages with the bacteria inside them and try to destroy them with the help of $T_H1$ lymphocytes.

## 2.7.- Activation and effects of T CD8 lymphocytes

In addition to the help that $T_H1$ cells provide to macrophages so that they can be activated to their full potential, despite the collateral damage this generates, $T_H1$ lymphocytes also help CD8 T cells reach their full killer potential. Remember that CD8 T lymphocytes are activated by dendritic cells that express peptides derived from virus proteins attached to MHC-1 molecules. These molecules bind peptides derived from proteins produced by the cells themselves (proteins not captured from the outside) and present them on the surface of dendritic cells.

Although we have ignored this question so far, this poses a serious problem, since for the dendritic cell to be able to present peptides in its MHC-1 molecules it should first be infected by a virus. This infection is what would lead the dendritic cell to produce internally the proteins of the virus, to degrade them to peptides in the proteasome, and to transport them to the endoplasmic reticulum, which is the organelle where they bind to the MHC-1 molecules before these and the peptides loaded in them are transported to the external membrane of the cell.

For dendritic cells to be infected by all viruses capable of affecting us, they should express the molecules on the membrane that allow viruses to bind to and enter the cells, or at least introduce their genetic material into them. In addition to the fact that this would mean that dendritic cells would always be expressing unnecessary proteins in the membrane just in case one virus or another could attack us, this would make the cells vulnerable to all viruses, which would certainly take advantage of this situation to contour the defense mechanisms, for example by quickly killing all infected dendritic cells, and thus preventing them from having time to travel to the lymph nodes to present the peptides derived from the virus proteins to the T cells.

Fortunately, we have a special type of dendritic cell that specializes in making an immunological exception. This exception is none other than the ability to present in the MHC-1 proteins captured from the outside. These dendritic cells are called **CD8α dendritic** cells, because in the case of the laboratory mouse (although not in humans), where they were discovered, they express this molecule on their surface. These cells are currently called **cDC1 dendritic cells**, whereas the classical dendritic cells that we have seen before are called **cDC2 dendritic cells.**

Let's take this opportunity to mention that the most common CD8 molecule, the one expressed by CD8 T lymphocytes, is formed by the union of two similar but different proteins, called **CD8α** and **CD8β**, each one produced by a different gene. While CD8 T lymphocytes express a **heterodimeric** CD8 molecule (i.e. formed by two units, from the Greek word *mero*, meaning unit, units that are different, hence the prefix *hetero*), the CD8 molecule of CD8 dendritic cells is formed by the union of two CD8α molecules, so in this case the mature molecule is a **homodimer** (formed by two identical protein units, hence the prefix homo).

Be that as it may, cDC1 dendritic cells are able to capture by macropinocytosis the viral particles found in the external environment without being infected by them. The particles are internalized into vesicles, as is always the case when proteins or viral particles are captured from the outside. These vesicles are nothing more than small cytoplasmic membrane spheres that contain floating particles present in the liquid captured from the outside. Strictly speaking, since they are in vesicles, the particles are not yet inside the cell, since they have not encountered the cytoplasm. For them to enter inside the cell, the vesicles must be dismantled, and their contents poured into the cytoplasm.

Well, cDC1 dendritic cells are not found in the skin or body surfaces, but reside in the lymphatic organs, where they arrive after being generated in the bone marrow, and where they are located waiting to capture the antigens transported to them by the lymph. These dendritic cells are different from those of the periphery and express co-stimulatory molecules for the correct activation of cytotoxic CD8 T cells that will be activated in the defense against viruses. These dendritic cells are also specialized in managing to transport the contents of the vesicles generated by pinocytosis to the interior of the cell. Once there, the proteins are treated enzymatically for their transport to the proteasome, where they will be degraded to peptides that will be treated in the same way as those coming from the proteins produced by the cell in the cytoplasm. These peptides will be transported to the endoplasmic reticulum where they will be charged to MHC-1 molecules and transported to the dendritic cell surface for presentation to naïve CD8 T cells. This process of uptake of proteins from the outside, their passage to the cytoplasm, their degradation to peptides and their presentation in MHC-1, instead of MHC-2, is called **cross-presentation**.

Dendritic cells, however, are very cautious and do not fully activate CD8 T cells unless other activated $T_H1$ cells allow them to do so. This implies that there is a communication mechanism, between at least three or even four different cell types, that must work properly to allow killer CD8 T cells to start killing virus-infected cells. These four cells are: a naïve CD8 T cell, a naïve CD4 T cell, a classic dendritic cell and a cDC1 dendritic cell. The reason that so many different cells are involved in the activation of CD8 cytotoxic T cells is that the cells that they are supposed to kill are from our own body, so this needs to be done for a good reason and with complete confidence that it is the right thing to do. Otherwise, the CD8 T cells would kill our own cells unnecessarily, which could even cause us to die.

Let's see what must happen for this communication mechanism between the four different cells mentioned above to work properly. Obviously, first, classical dendritic cells must capture viral particles from the exterior. These particles usually come from a viral infection that is already taking place and that could not be avoided by other protection mechanisms, for example, by virus-neutralizing antibodies. The classical dendritic cells with the captured virus particles will travel through the lymphatic system to the lymph nodes near the infection where they will present the peptides derived from the captured viruses in their MHC-2 molecules. This implies that these cells can only activate naïve CD4 T cells, since they do not present virus peptides in their MHC-1 molecules. Fortunately, once they reach the node, classical dendritic cells will also transfer, by mechanisms not yet clear, the viral particles to cDC1 dendritic cells that are resident in the local lymph nodes. These cells will capture the viral particles from classical dendritic cells, and also particles that can be found floating in the lymph, and will generate peptides derived from them and present them in both MHC-1 and MHC-2 molecules. These cells will therefore be able to activate both CD4 and CD8 T cells. Thus, we have two classes of dendritic cells, classical and cDC1, capable of activating CD4 T cells. When these cells find a virus-derived peptide linked to an MHC that they recognize with their receptors, they will be induced to mature and differentiate into $T_H1$ cells by the action of cytokines secreted by dendritic cells. On the other hand, cDC1 dendritic cells will also be able to activate naïve CD8 T cells to become cytotoxic CD8 T cells, but this activation will not occur properly unless the $T_H1$ cells have been activated first. These $T_H1$ cells therefore detect peptides of the same virus with their receptors, but not the same peptides as those detected by the CD8 T cells. The reason why the peptides are not the same is easy to understand. As we have already said, $T_H1$ lymphocytes are CD4 T cells, which detect peptides attached to MHC-2 molecules, but CD8 T lymphocytes detect peptides attached to MHC-1 molecules. Both types of molecules bind different peptides, so the viral peptide or peptides detected by CD8 T cells are necessarily different from those detected by $T_H1$ CD4 T cells, even though they come from proteins of the same virus.

It is useful here to briefly mention that some dendritic cells can be infected by viruses, such as Herpes simplex or flu viruses, and be killed by them before they can present antigens to the T cells. This makes it difficult to mount an adaptive immune response to these viruses. Fortunately, the infected dendritic cells can activate themselves by detecting virus components with their Toll-like receptors and travel to the lymph nodes. Once there, in a dying state, they can transfer the viral antigens captured or generated in the infection to the cDC1 dendritic cells resident in the lymph nodes.

Thus, we have dendritic cells that present viral peptides in both types of MHC molecules. When this happens it means that, without a doubt, a viral infection is under way and it is necessary to stop it. The remedy is to generate neutralizing antibodies against the virus to prevent it from infecting more cells, and to induce the death of the infected cells as quickly as possible, since each infected cell is a virus factory that will release hundreds or thousands of new virus particles to the exterior. If unchecked, they will spread in geometric proportion even faster than the bacteria.

Dendritic cells thus present peptides to both naïve CD4 and CD8 T cells. The naïve CD4 T lymphocytes are activated and become $T_H1$ lymphocytes, which will do two things. The first is to secrete cytokines that activate the CD8 T cells and stimulate their growth. Among these cytokines is **interleukin-2 (IL-2)**, one of the first to be discovered and **one of the most important cytokines for T cell activation**, since it induces proliferation and the generation of clones of thousands of activated T cells. The second action performed by $T_H1$ cells is to further activate the virus-activated dendritic cells. This extra activation results in the expression of a greater number of co-stimulatory molecules, mainly B7-1 and B7-2 (although there are also others), on the surface of the dendritic cells. This higher concentration of co-stimulatory molecules is necessary for the complete activation of CD8 T cells, which once they are activated to their full potential leave the lymph nodes in search of virus-infected cells to kill them. This search is carried out by the same mechanisms of adhesion and penetration of the endothelium that we have already explained, which occurs at the sites of inflammation and infection.

Cell death induced by these lymphocytes is also a process of suicide, of apoptosis, by means of molecules that lead to the perforation of the cell membrane of the infected cell. The molecules that produce pores in the infected cell that an activated CD8 T lymphocyte encounters are called **perforin** and **granulysin**. These molecules act by causing the formation of pores in the infected cell membrane through which other molecules produced and secreted by CD8 T cells can enter: **granzymes**. These are the proteins possessing the enzymatic activity capable of initiating apoptosis, but to initiate it they must penetrate inside the infected cell, which must also present bound to its MHC-1 peptides of the virus that the lymphocyte T CD8 must recognize as foreign. As far as we know, once the activated CD8 T lymphocyte has recognized its antigen on the target cell, the signal it receives thanks to this recognition allows it to increase the strength with which integrins can adhere to the adhesion molecules that the target cell expresses on the membrane. The CD8 T lymphocyte thus adheres strongly to the target cell, but cannot adhere to other cells that do not display peptides that its T receptor recognizes as foreign. Once strongly bound, the biochemical signal received via the T receptor triggers the secretion of its granules (which are also small intracellular vesicles), in which perforin, granulysin and granzymes are stored. These molecules are secreted bound together with yet another molecule, called **serglycine**, which is the main proteoglycan (protein and carbohydrate complex) of the CD8 T cell cytotoxic granules, and acts as a transporter. After recognition by a cytotoxic CD8 T cell of an antigen on the surface of a virus-infected cell and the adhesion of both cells, the lymphocyte releases the contents of its cytotoxic granules into the contact zone of the CD8 T cell and the infected cell, i.e. the contents of the lymphocyte granules are released in a highly concentrated manner over a specific region of the target cell membrane. Perforin and granulysin allow the formation of pores in the target cell membrane and direct the entry of the granule contents into its cytosol. Once inside this cell, the granzymes then act on other molecules located in the cytosol of the target cell. One of them is the so-called **BID**. Another is called **pro-caspase 3**. The action of granzymes on these proteins triggers the process of apoptosis. Once the CD8 T lymphocyte has induced apoptosis in the target cell, this cell stops sending signals to the T receptor, as the MHC-1 molecules are degraded, allowing the CD8 T lymphocyte to detach from it and glide thanks to the movement of body fluids in search of other infected target cells. It has thus been calculated that a single activated CD8 T cell can kill up to a thousand infected cells before dying of exhaustion. The killing efficiency of these lymphocytes is therefore very high, hence the need for these lymphocytes not to be activated frivolously, while at the same time there are regulatory mechanisms to curb their activity once the infection has been overcome. However, it is important to mention that whereas co-stimulation of CD8 T cells is necessary for their activation, co-stimulation is not necessary for the killer activity of CD8 T cells. This seems quite logical since infected cells are not usually dendritic or other antigen-presenting cells, which are the only ones capable of expressing co-stimulatory molecules on their membrane. If the cells of the organism needed to express co-stimulatory molecules to allow the activity of CD8 T cells, all the cells of the organism could stimulate naïve CD8 T cells against their own antigens, which would dramatically increase the chance of developing autoimmune diseases. For this reason, throughout evolution, the antigen presentation function has been delegated only to "professional" cells for this work, which are the dendritic cells.

## 2.8.- $T_H17$ LYMPHOCYTES

We have seen how $T_H1$ cells provide an important bridge between the cells of innate immunity, particularly macrophages, and adaptive immunity. The combination of the two is essential to control the threat posed by bacteria that have become established in a point of infection.

However, macrophages are not the only cells of innate immunity involved in infection control. As we know, neutrophils are also key phagocytic cells. They are the first to arrive at the source of infection and once there they become completely involved in the fight against bacteria using fascinating mechanisms that we will describe later.

As in the case of macrophages and $T_H1$ helper cells, neutrophils also work with a type of helper T cell that enhances their activity. This helper T cell is the **$T_H17$** cell.

$T_H17$ cells are the first to be activated in response to a bacterial infection. This is probably because their role enhances that of neutrophils, which are the main phagocytic cells against extracellular bacteria. $T_H17$ cells are derived from naïve CD4 T cells that have found an antigen presented by dendritic cells that stimulate them with certain cytokines. In this case, the inducing cytokines are **TGF-β** (Transforming Growth Factor beta), **IL-6**, **IL-21** and **IL-23**. These cytokines are produced by dendritic cells that have been activated by antigens found in extracellular bacteria and fungi.

The role of $T_H17$ cells is important in controlling pathogens that can enter the body via the mucosal surfaces, i.e. the intestine, mouth, airways and urogenital tract. $T_H17$ cells, once activated, will produce the cytokines **IL-17A** and **IL-17F**. These cytokines act on skin cells and mucosal surfaces and incorporate them into the fight against microorganisms. First, skin cells activated by IL-17A and IL-17F will secrete chemokines that will attract neutrophils to the site of infection. However, we already know that epithelial surfaces are an important barrier against infection, and they are an active barrier. Remember the poisonous walls of the castle. Well, these walls can be activated by these two cytokines to increase their capacity to secrete bacterial poisons and to produce other cytokines that travel from the site of infection to the bone marrow and stimulate the generation of more neutrophils from the stem cells. The release of these cytokines by the skin cells and not by the attracted neutrophils is a way to distribute the defensive effort more efficiently among the different cells of the body. In this way, the neutrophils concentrate on their actual defensive task, while leaving it to other cells to produce signals that send information to headquarters, i.e. to the bone marrow, about what kind of reinforcements are needed.

$T_H17$ cells also secrete the cytokines IL-21 and IL-23, which will enhance the generation of more $T_H17$ cells from naïve CD4 T cells. Thus, the initial generation of $T_H17$ cells promotes the activation of more cells of this class, to the detriment of the generation of different types of activated CD4 T cells, such as the $T_H1$ cells we have seen before, but also other types of CD4 T cells that we will see later.

This reflects an important feature of the immune system. This is that once it has made a decision about the type of defense mechanism it should be used to cope with a threat this mechanism is maintained as long as the threat persists, and alternative mechanisms, which would normally not be effective or may not yet be needed, are inhibited. Only if these initial defense mechanisms are not sufficient to contain the infection can the immune system boost other mechanisms that join or replace the previous ones in the fight.

Thus, $T_H17$ cells will be activated first and will be located on mucosal surfaces and bacterial infection sites. The secreted IL-17 cytokines activate the epithelial cells themselves and induce them to secrete antibacterial substances and neutrophil-attracting chemokines. It is possible that the antibacterial substances and the attracted neutrophils can keep the infection at bay on their own, with no other help than the complement system. If that happens, no more resources need to be devoted to containing the threat. However, if the infection persists, it is possible that dendritic cells that continue to carry antigens from the source of infection to the lymph nodes will induce differentiation to other types of T cells which, in this case, will enhance antibody generation or boost macrophage activity. In fact, $T_H17$ cells can be converted, if necessary, into $T_H1$ cells if they are stimulated by the cytokine IL-12 secreted by dendritic cells. All these mechanisms together, i.e. antibacterial substances, the complement system, neutrophils, macrophages and antibodies, which will also help macrophages to phagocyte the bacteria and to neutralize and opsonize the viruses, will probably eradicate the infection.

In any case, the immune system needs to be able to put in place the right defense mechanisms by achieving a balance that maximizes efficiency and minimizes the consumption of unnecessary resources for defense. It is not an easy task, but understanding that this is what the immune system must achieve to guarantee the survival of the organisms can help us to accept the reasons for the complexity of this system, its mechanisms for communicating information between the different cells, for making decisions in accordance with that information, and for the different mechanisms of defensive action.

Returning to the $T_H17$ cells, we can ask ourselves why they are generated first? Why is it important to attract neutrophils to fight bacteria first, rather than triggering other, perhaps more sophisticated, mechanisms of adaptive immunity? The answer to this question lies in the impressive properties of neutrophils and their extraordinary effectiveness in fighting bacteria.

### 2.8.1.- The Formidable Neutrophils

Neutrophils are so called because they are neutral in terms of the chemical properties of the coloring agents traditionally used to identify different blood cells. Neutrophils are the most abundant leukocytes in the blood, constituting between forty and seventy percent of these. Like all leukocytes, neutrophils are generated in the bone marrow from stem cells, from where they exit to the blood and circulate through it. When neutrophils reach the capillaries near a site of infection, they attach to them thanks to the interaction of their sialil-Lewis$^X$ with E and P selectins of these, and they carry out the extravasation process mentioned above. Neutrophils are the cells with the greatest capacity for adhesion to the endothelium, compared to other leukocytes, and can adhere and roll over it pushed by the blood flow even when the force of the latter would prevent other leukocytes from doing the same. The reason for this great adhesive capacity of neutrophils to the activated endothelium is that they have extensions of the membrane that act as elastic bands for retention. When these extensions (called, in fact, slings) adhere to the endothelium and the neutrophil begins to roll on it, the extensions stretch by rolling on the surface of the neutrophil, thus slowing down its rolling. This allows the neutrophils to bind tightly to the endothelium and pass through it, even in places where this is impossible for other leukocytes. Once out of the blood capillaries, they are directed to the source of infection by following the concentration gradient of chemokines generated by macrophages and sentinel cells that were already at site through which the bacteria have penetrated.

Until 2004, neutrophils were known to perform two fundamental functions. The first was the phagocytosis and digestion of bacteria; the second was the secretion of antibiotic substances. To perform the first function, neutrophils have receptors capable of detecting some substances that bacteria secrete to their environment as a result of their metabolism. The detection of these substances causes complex changes in the cytoskeleton of neutrophils that prompt them to change their shape in a plastic way and generate pseudopodia to "swim" in search of the bacteria, which they end up trapping, phagocyting and digesting.

Neutrophils also have granules in their cytoplasm that are loaded with substances that are toxic to bacteria. These granules are classified into two classes, **primary granules** and **secondary granules**, according to the class of molecules they both contain. The primary granules contain various antimicrobial substances, such as defensins and cathelicidins **(section 2.1)** that generate pores in the bacterial membrane. They also contain some enzymes that can digest both bacteria and our own tissues, especially **neutrophil elastase**. The primary granules fuse with the secondary granules and phagosomes, leading to the death of the bacteria the latter contain. The content of the primary granules can also be released to the outside and the enzymes can thus act on the proteins of the extracellular matrix that maintain the integrity of the tissues, degrading them and thus facilitating the navigation of the neutrophils through the tissues in search of the bacteria at the source of infection.

The secondary granules can also fuse with the phagosomes, along with the primary ones, helping the anti-bacterial activity of the latter, but their contents can also be released to the exterior, since these granules contain substances that contribute to slowing down bacterial reproduction, such as **lactoferrin**, another protein that sequesters iron and prevents bacteria from capturing it and thus allowing it to grow. In addition, a very important protein contained in these granules is called **properdine**. This protein acts by promoting the activation of the complement system on the surface of the bacteria, which helps their opsonization and phagocytosis. When some substances produced by the bacteria and certain cytokines are detected, the neutrophils release the content of these granules to the exterior, thus helping to overcome the infection.

In 2004, a third function of neutrophils was discovered, which is truly impressive, as it was confirmed that neutrophils are also capable of secreting no less than strands of DNA. The secreted DNA has anti-microbial substances attached to it and, at the same time, because it is a long molecule, it forms a molecular grid that traps bacteria and immobilizes them. These molecular DNA nets have been called **neutrophil extracellular traps** (which generates the appropriate acronym NET). There are three ways that neutrophils can secrete this DNA. One way leads to cell death in a kamikaze act, but not the other two. The DNA can be nuclear or mitochondrial.

The secreted DNA traps the bacteria in the DNA net and the antibiotic substances concentrated there effectively kill the bacteria. At the same time, the net prevents those bacteria that have been able to escape the action of the antibiotic substances from dispersing throughout the body and facilitates the phagocytic action of macrophages. Macrophages, like cell spiders, are attracted to neutrophils and use the DNA net to better eat their immobilized prey. A final advantage of the molecular DNA nets is that they also immobilize the antibiotic substances produced by the neutrophils, which can be toxic to our own cells if they are dispersed throughout the body.

These incredible properties of neutrophils make them very effective cells for antibacterial fight, so if adaptive immunity is to be put into action, generating cells like $T_H17$ first, capable of attracting them to the source of infection and boosting the activity of neutrophils, is a sensible strategy to slow down the infection immediately and wipe it out as soon as possible, dedicating the defensive resources in a very effective way. Only when neutrophils, even with the help of $T_H17$ cells, are overwhelmed and cannot eradicate the bacterial infection, other mechanisms are set in motion involving the generation of other types of adaptive immunity cells, specialized in the activation of macrophages to their maximum potential, as we have seen, or in the production of antibodies, as we will see below.

## 2.9.- $T_{FH}$ Lymphocytes

We have already commented before that many bacteria live in the extracellular spaces, that is, those that reside between the cells. These bacteria, instead of allowing themselves to be phagocyted to live inside macrophages, try to escape phagocytosis by coating themselves with molecules that protect them from detection by phagocytes, or even by complement molecules. Viruses, likewise, although they reproduce inside the cells, need to leave the cells once they have been produced in order to infect other cells and, therefore, part of their life cycle takes place outside the cells.

This outer world that extends between our cells and is also found in the fluids of our body (the so-called humors) is the battleground where complement and antibodies, which we spoke about a few pages ago, contribute to the fight. Antibody molecules are fascinating because, although they all have a very similar structure, they are capable of binding to practically any molecule that exists in Nature, and even to molecules that do not yet exist but will exist in the future, such as, for example, a drug that has not yet been invented. This statement may be very surprising, but it is rigorously true. I promise you that later I will explain to you how this is possible. Suffice it to say now that the immune system, thanks to antibodies, has found a way to anticipate the potential threats from the outside world, and defend against them.

The antibodies are the central molecules of the so-called **humoral immune response**, because it is the one that develops in the humors and involves molecules as executors of the main defense activity, unlike the **cellular immune response**, which always involves the action of certain cells, such as macrophages or $T_H1$ cells, as we have seen. The humoral immune response depends, however, like all the functions of our body, on the functioning of certain cells, in this case, the cells that produce the complement and acute phase molecules (the liver) and the antibodies (the B lymphocytes). Only the latter cells can produce antibodies. In the absence of such lymphocytes, people or animals lack antibodies and suffer from severe infections by bacteria and viruses. However, for the correct generation of most antibodies by B lymphocytes, the collaboration of T lymphocytes, called $T_{FH}$ lymphocytes, is necessary. In the absence of these, the generation of antibodies is seriously affected, which also causes severe immunodeficiency.

Before we get into the role that these lymphocytes play in the generation of antibodies, let's talk briefly about the function of these molecules and why they are so important for keeping us healthy. We have already said that antibodies are molecules generated in response to the detection of a foreign molecule by one or another B lymphocyte. These, just as T lymphocytes have different T-cell receptors on their surface, have so-called B-cell receptors, which are also different among different B lymphocytes. We will see later how B lymphocytes achieve this, but it is scientifically true that each B lymphocyte shows on its surface thousands of molecules of a particular receptor, molecules that are identical to each other in a B lymphocyte itself, but different from the receptor molecules of other B lymphocytes. Generally, we all therefore possess billions of B lymphocytes, each with the capacity to produce a specific receptor molecule that is different from those of other B lymphocytes.

An antibody to a foreign substance is only produced if a B lymphocyte can detect that foreign substance with its receptors. Normally only a few of the billions of B lymphocytes we have will detect the molecules of the foreign substance with their receptors when they bind to some specific region of the surface of the foreign substance. Notice this idea is important. The molecules of different substances are actually objects with a certain shape. Just as a chair has one shape and a sofa has another, so do the molecules. B-lymphocyte receptors are also molecules, and some of them may have a shape that is spatially complementary to that of some part of the surface of a foreign substance. Making use of the chair simile, and reducing it to the size of a molecule, some B-lymphocyte receptors will possess a shape that is complementary to that of the leg, another to that of the seat, and another to that of the back. The receptors of these lymphocytes will all be able to detect and bind to the chair, although they do so in different places. These different places of the molecule surfaces are called **epitopes**, a word derived from the Greek words *epi*, meaning surface, over, and *topos*, meaning place.

The detection of an epitope by the receptors of a concrete B lymphocyte leads to the activation of the lymphocyte, which stimulates its division and the generation of a clone of initially identical cells that secrete identical antibodies able to bind to the detected epitope. However, without the help of $T_{FH}$ cells, B cells can only secrete one class of antibody, IgM, out of the five classes that exist. These classes are named **IgD**, **IgM**, **IgA**, **IgG** and **IgE**. The classes are also called **isotypes**. The letters Ig are an abbreviation of the word "immunoglobulins", which is the generic name by which the antibodies are called.

Why are there five different kinds of antibodies? The answer is simple: each class performs a different function. Each class is adapted to communicate, to different types of cells or to the complement system, the information that they have detected one of the epitopes of a given antigen. Depending on the cells that receive this information, different effector mechanisms will be set in motion, specialized in trying to kill the enemy with maximum efficiency.

How does a B lymphocyte that has detected an epitope of a foreign substance know what kind of antibody to produce? This information and command must be received from a $T_{FH}$ lymphocyte that has been activated by a dendritic cell in the lymph node. The dendritic cell must direct the activation of a naive T cell to a $T_{FH}$ lymphocyte according to the information the dendritic cell has detected about the nature of the microorganism.

Here again we find the same kind of mechanisms that we have already found in the activation of $T_H1$ cells. The dendritic cell has detected an enemy through one or more of its Toll-like receptors and, because of the nature of the substances inherent to this enemy, it has been activated. This leads it to express co-stimulatory molecules on the membrane and to secrete certain cytokines. Both the co-stimulatory molecules, (especially one, called **ICOS**), and the cytokines secreted by the dendritic cell, induce the naïve CD4 T cell, which recognizes with its receptor a foreign peptide presented by an MHC-2 molecule on the dendritic cell membrane, to mature and differentiate to a $T_{FH}$ cell so that this one reproduces and generates a clone of thousands of identical $T_{FH}$ cells.

Once mature, the $T_{FH}$ cells will remain in the lymph node, where they will interact with the B cells, but only with those B cells that have detected some epitope of the same antigen that the $T_{FH}$ cells have detected in the form, as we have said, of a peptide attached to an MHC-2 molecule on the surface of the dendritic cell that activated the naive CD4 T cell from which they are derived. A similar situation has been found before with the macrophage and the $T_H1$ cell. In this case, however, it is the B lymphocyte that must present the $T_{FH}$ cell with the same peptide attached to an MHC-2 molecule that the dendritic cell has presented to the naive CD4 T cell to activate it and generate the $T_{FH}$ cell clone. If a $T_{FH}$ cell detects on the surface of the B lymphocyte the peptide that activated its "mother" (the naïve CD4 T cell from which the "daughter" $T_{FH}$ cells are generated), it will secrete certain cytokines that will stimulate the B cell to secrete the most appropriate class of antibody to fight the type of threat that the dendritic cell has detected in the first place.

In this way, the cells communicate with each other and transmit various information. Firstly, the B cell that has identified an epitope, captures it, internalizes it attached to its B-cell receptor and digests it into several peptides that it will present attached to MHC-2 molecules. At least one of them can be recognized by a $T_{FH}$ cell activated by a dendritic cell that presented to it the same peptide that the B lymphocyte now presents. In this way, the $T_{FH}$ cell knows that the B lymphocyte has detected the same antigen that was initially detected by the dendritic cell in the source of infection and that this cell presented it to its "mother" (the naïve CD4 T cell from which the "daughter" $T_{FH}$ cells are originated).

Secondly, the $T_{FH}$ cell must communicate to the B lymphocyte that it has been identified and must confirm the information that this lymphocyte has indeed detected an epitope belonging to an antigen of a potentially dangerous microorganism and, therefore, it must divide to generate a clone of activated B cells. These are the ones that will produce antibodies of the right kind to fight it. The $T_{FH}$ cell communicates this information to the B cell in two ways. The first is like that already seen with macrophages and involves the molecule CD40L, expressed by the

$T_{FH}$ cell on its surface, which will interact with the CD40 receptor, expressed by the B cell. This will send to the B cell nucleus a first signal to activate certain genes. The second way in which the $T_{FH}$ cell communicates information and gives commands to the B lymphocyte is through the secretion of specific cytokines, particularly those called **IL-21** and **IL-4**. The combination of these two signals (co-stimulatory molecules and cytokines) causes B cells to synthesize and secrete the class of antibody best suited to deal with the threat first detected by the dendritic cell.

### 2.9.1.- ANTIBODY CLASSES

When a B cell recognizes an antigen epitope, depending on the type of signals it receives from the $T_{FH}$ cell, it will produce and secrete one antibody isotype or another. However, it is important to consider that whatever class of antibody is produced, it will always recognize the same epitope, i.e. the same external region of a given antigen. Therefore, class switch in a given immunoglobulin does not mean that the antigen is recognized and detected more efficiently, but it does mean that it is fought more efficiently. The mechanisms of combat set in motion are different depending on the class of antibody that the $T_{FH}$ cells, (and some other types of T cells that we will discuss later) communicate to the B cells that they must produce. Depending on the isotype, the antibodies will communicate the information only to specific molecules and cells of the immune system that they have bound to an antigen. Of course, when we talk about information communicated between cells or between molecules, we already know that we are talking about molecular interactions between ligands and receptors. Which ligands and receptors are we talking about here?

To understand why different receptors and molecules interact with different classes of antibodies, we have to look briefly at the structure of these molecules, which, despite being of different classes, show a similar structure. The basic part of antibody molecules is formed by the binding of two identical couples of two different protein chains. Two of the chains that are identical to each other are bigger and are therefore called the **heavy chains**. The other two smaller identical chains are called the **light chains**. To form a mature antibody molecule, a heavy chain must be joined to a light chain, and then these two joined chains must be joined to another equal pair, formed by the joining of and identical light chain and an identical heavy chain. So, the antibody is formed by the union of two heavy chains and two light chains which are identical to each other, although always a heavy chain is first joined to a light one, and then this combination of two chains is joined to another unit formed by the same combination. **The antibody molecule is therefore said to be a tetramer**, which is formed by the union of two heterodimers.

The four protein chains form a Y-shaped molecule. The arms of the Y are made up of a part of the heavy chain and the entire light chain. The ends of these arms detect the antigen. For that reason, the Y arms are called **Fab** (antigen binding fragment). The detection of the epitope on the surface of the antigen is carried out by both a specific area of the light chain and an equivalent area of the heavy chain, which are located at the ends of the Y-arms. On the other hand, the "tail" of the Y is formed only by the union of the terminal ends of the two heavy chains. This "tail" is what differentiates the different classes of antibodies and gives them different effector properties, according to the cells and molecules this part interacts with. The scientific name of this "tail" is **Fc** (crystallizable fragment), since it can be crystallized with some ease from solution.

It is very important to know the fact that the antibody Fc region will work as a ligand for other receptors or molecules present in the cells or in the external fluids. The binding to the Fc region of these molecules or receptors present on various immune cells is what gives the different classes of antibodies their different defense properties, i.e. their specific role as weapons against microorganisms.

The receptors present on the surface of some cells of the immune system that bind to the Fc regions of the antibodies are called, as we have said, Fc receptors. The soluble molecules that can bind to the Fc region of the antibodies found in the external fluid that bathes the cells are the first molecules of the complement which, as we have seen, actively participates in eliminating bacteria by opsonization and generating pores in their membrane. The following figure represents the structure of a typical antibody molecule.

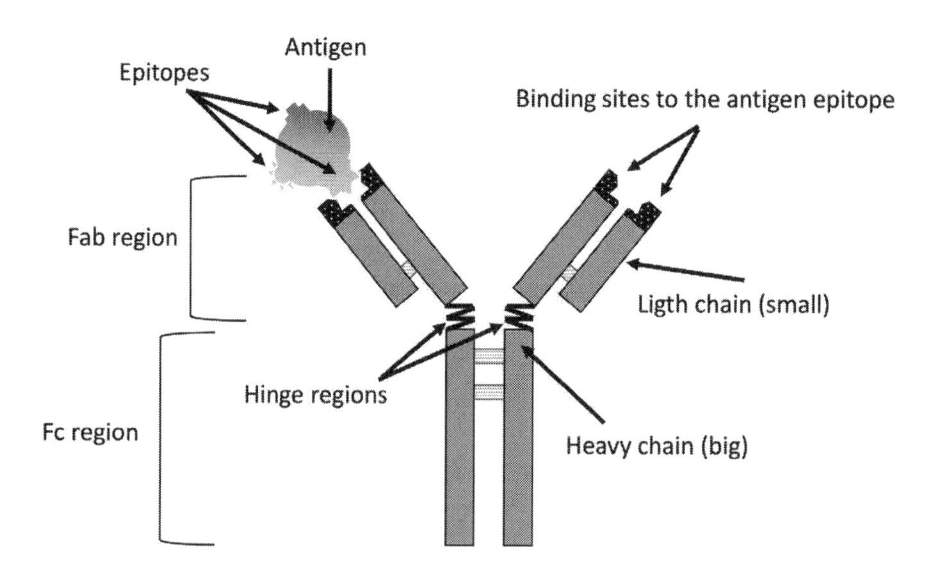


***Representation of the antibody molecule structure***

For example, the first class of antibody produced and secreted to the exterior is IgM. This antibody is formed by the binding of five basic units of immunoglobulin, i.e. five Y's. This structure gives it certain properties that other immunoglobulins with a different structure and different heavy chain do not possess, and that is because once bound to an antigen, it displays five Fc regions in a precise conformation. This conformation of the five Fc regions allows IgMs to act as ligands for the first molecule of the complement system and to activate this system immediately just on the surface of the antigen they have bound to. Complement can then act more effectively.

The secretion of immunoglobulin classes other than IgM requires the process of class switch referred to above. When the B cell senses that the $T_{FH}$ cell expresses CD40L, as well as certain cytokines, it can stop producing IgM and start producing IgA, IgG, or IgE, according to the cytokines detected. IgD is secreted only in very small amounts, it is not known why. Each of the three major classes of immunoglobulins, IgA, IgG, or IgE, has receptors for their Fc regions expressed on different cells. For instance, macrophages have receptors for IgG Fc regions. When a macrophage or a neutrophil detects several of these molecules bound to the same antigen, it captures the bacteria and the signal emitted by the

Fc receptors induces the mechanism of phagocytosis. Thus, the presence of immunoglobulins secreted by B cells helps the defense activity of macrophages and neutrophils in the focus of infection.

It may be necessary to clarify that not all antigen-activated B cells change the class of immunoglobulins they produce, since different classes of antibodies to a given antigen coexist in blood plasma and tissue fluids. The proportion of B cells resulting from the clonal expansion of a B lymphocyte initially activated by an antigen epitope that change the class of the antibody depends on how many find a $T_{FH}$ cell, among other possible classes of T cells, to induce that change. Eventually it is possible for all B cells to change the antibody class, but this is a gradual process. The half-life, or speed of degradation, of each class of antibody produced also affects its concentration in the blood and body fluids.

Normally, there is simultaneous binding of several IgGs to the same antigen, and this for several reasons. Firstly, the amount of antibody secreted by the activated B cells is abundant, so there is usually enough antibody to bind all the antigens. Second, it is also common for a microorganism, for example, a bacterium, to have the same epitope repeated several times on its surface, to which antibodies directed against produced by a single clone of B cells will be directed, since each B cell produces a different antibody directed against a particular epitope. In this case, the bacterium will have several molecules of these IgGs attached to its surface. Finally, it is also usual that several B cells detect a different epitope of the same antigen and produce different IgGs that will bind to this antigen through the epitope they detect. Thus, bacteria or viruses against which antibodies have been generated will be coated with several of these molecules, not far apart from each other.

The binding of several IgG molecules to the antigen at the same time is an important phenomenon that communicates information to macrophages, to phagocytes in general, and to complement and some of the acute phase molecules, that they have encountered an antigen they must phagocytize, or that complement must be activated. This is important because, let us consider: what would happen if the Fc receptors of phagocytes or complement or acute phase molecules could bind to the Fc regions of free IgGs, without these having previously bound the antigen? Obviously, if this were to happen, phagocytes would be able to phagocytize free IgG molecules without antigen binding to them, which would lead to a waste of energy and resources, and would even be dangerous, since free IgG would confuse the phagocytes into thinking that they had found an antigen, and this would diminish the effectiveness of their defensive function. So, it is important that the phagocytes know how to distinguish whether, when they detect an IgG with its Fc receptor, it is bound to an antigen or not. How do they know that?

The way Nature has managed to get this information to the phagocytes is by generating receptors for the Immunoglobulin Fc regions that are not strong enough to hold together only one Fc of these molecules. It is not that a free IgG cannot bind to this receptor on a phagocyte membrane, but if it does bind to it, the receptor cannot retain the immunoglobulin molecule, and it is released. On the other hand, if several Fc receptors on the phagocyte membrane bind to several Fc immunoglobulin regions at the same time, which can only happen if the antibodies are bound to an antigen simultaneously, the Fc receptors add up their binding strength and can retain attached immunoglobulins. All these simultaneous bonds cooperate with each other and now prevent the phagocyte from releasing the antigen that carries the bound antibodies, as it has been captured by numerous Fc receptors that held it captive. In addition, the IgG Fc receptors expressed by the phagocyte, which are normally moving aimlessly on its membrane as if they were drifting boats on the surface of a lake, now get in proximity to each other as they bind to the antibodies bound to the antigen. This proximity is necessary for them to set in motion the molecular mechanisms within the cell that will lead to the phagocytosis of the detected antigen. In this way, phagocytes can know that they have detected an antigen that must be phagocyted, and they never phagocytize free IgGs.

The antibody isotype that is produced in the greatest quantity is IgA. In fact, B cells produce more IgA than all other antibody classes combined. Most IgA is produced by B lymphocytes that are located near or on the mucosal surfaces of epithelial tissues. The **mucosal surfaces**, as their name suggests, produce mucus, a normally sticky substance that binds numerous bacteria and microorganisms and is secreted to the exterior, helping to prevent those microorganisms from entering our bodies, where they could cause an infection. As mentioned above, mucus contains proteins called **mucins**, which carry a large amount of carbohydrates, which is the main reason for the sticky nature of mucus. However, secreted mucus is not enough to control the growth of microorganisms on well nourished, moist surfaces, such as the surface of our intestines. To achieve this control, large amounts of antibodies need to be secreted into the gut. This antibody is of the IgA isotype. In the intestine, IgA binds to the surface of the bacteria, coats them, and prevents them from adhering to the cells of the intestinal epithelium, which is a necessary condition for later penetration into the body and an attempt to establish a source of infection. The peristaltic movements and fluids that continuously pass through the intestine thus expel many bacteria linked to IgAs with the feces. In this way, the number of bacteria in the intestinal flora is controlled and species of bacteria that could be pathogenic are eliminated from it.

The five classes of antibodies thus provide a variety of mechanisms for fighting microorganisms, depending on their nature and location (skin, intestine, etc.). However, in any case, the class of antibody that is produced and secreted first is IgM, and it is worth asking why this is so. Why has Nature, being able to produce as an initial response to a threat an immunoglobulin consisting of only one Y, which would be easier, chosen to produce an immunoglobulin consisting of five Ys first? As with almost everything that Nature does, there is a good reason, which I will try to explain below.

We have previously explained that B lymphocytes possess receptors capable of binding to any molecule in the outside world, even molecules that do not yet exist, but which can be synthesized by a laboratory in the future. This is strictly true, however surprising it may seem. As I also mentioned before, we will explain this later, and you will understand how this is possible. It is enough now to know that each B lymphocyte, during its maturation, has generated a receptor against some foreign substance to the organism that the lymphocyte does not know what it is, but that, if by chance it encounters it during its lifetime, it will bind to it and the lymphocyte will be activated and will generate antibodies that will bind to that substance. These antibodies, in case it was not clear before, are in fact the same receptor molecules for the foreign substances, but they have been slightly modified to be secreted to the external environment, instead of being placed on the cell membrane.

However, while the B-lymphocyte receptor repertoire is virtually unlimited with respect to the diversity of substances it can detect, it is also true that when a receptor detects a foreign substance, it does not always do so with the same strength regardless of what that substance is. In other words, Nature has been "intelligent" enough to generate receptor molecules against virtually anything, but it cannot always generate them in such a way that they bind to that thing with enough strength.

This lack of strength with which most of the receptors of the B lymphocyte repertoire bind to external substances has been able to be compensated, however, by allowing the binding not of a single receptor to an antigen, but of several of them to the same antigen and at the same time. Only when this happens, the antigen is retained for a sufficient time by the B lymphocyte, and sufficient membrane receptors gather at the site of binding to this antigen. All these receptors together now make it possible to trigger the molecular mechanisms necessary to transmit, in the form of activated molecules, the signal to the cell nucleus, which allows the genes necessary for the activation of the B lymphocyte to be set in motion and for the B lymphocyte to produce and secrete the antibodies.

This activation is, therefore, allowed by the binding to several identical, or very similar, epitopes to which many B-lymphocyte receptors bind with little force. However, since there are many receptors bound, the antigen with its epitopes remains bound by all of them to the B lymphocyte and is internalized and digested by it for the presentation of protein fragments by MHC-2 molecules. These fragments are the ones that the $T_{FH}$ cells we have seen before will detect.

After its activation, the B lymphocyte will produce antibodies that bind to the same epitope to which its receptor initially bound, since as we have said, the antibodies are nothing more than receptors modified to be secreted. This means that the antibodies will bind with little force to that epitope. Since the basic antibody molecule is Y-shaped, it has only two antigen binding sites at the ends of the Y-arms. Even binding two epitopes at the same time does not provide sufficient binding force to keep the antibody bound. The antibody, in the aqueous medium in which it must be to act, is receiving, every second, millions of blows and pushes from water molecules, other protein molecules, etc. Therefore, if the antibody does not bind strongly enough, it is separated as a result of that molecular agitation proper to the aqueous medium necessary for life.

This indicates that, if the antibodies produced were only single molecules, most of them could not bind to their antigen long enough to neutralize it. To enable sustained binding, Nature has, throughout evolution, bound up to five antibody molecules together to form IgM, which is the first class of immunoglobulin to be produced. This IgM thus has ten binding sites to the same antigen, and these ten binding sites are usually enough for IgM to remain bound to it. This cooperation of several antigen-binding sites to keep the antibody attached to the antigen is called **avidity**. Avidity depends on the number of binding sites that one substance has for another, in this case antigen to antibody. On the other hand, the strength with which each binding site binds is called **affinity**. Thus, IgMs are high avidity antibodies, but usually low affinity ones (some may, however, have a high affinity purely by chance, if the receptor of the B lymphocyte that produces IgM has detected an epitope to which it binds strongly). Each binding point binds weakly to the epitope, but because union means strength, the binding of ten such interaction points to ten identical epitopes keeps the IgM fixed to its antigen. The binding of IgM molecules to antigens, as we have also explained above, can now trigger activation of complement, which will favor phagocytosis of the microorganism by phagocytic cells.

This mechanism of complement activation has also been favored by natural evolution, since complement activation results in an activation of the overall immune response. The activation of complement by IgM, and the stimulation of antigen phagocytosis by dendritic cells and macrophages that this entails, favors its digestion and presentation in the form of peptides bound to the MHC-2molecules of these antigen-presenting cells which, from the point of infection, travel to the lymph nodes via the lymph, as explained above. Thus, the initial activation of complement by IgM stimulates the inflammatory response and the generation of a greater diversity of antibodies directed against as many antigen epitopes as the antigen possesses.

So, there is a good reason why IgM is the first immunoglobulin to be produced. Fortunately, however, it is not the only kind that is produced, or we would otherwise have serious problems controlling certain bacterial infections. Remember that once the B lymphocyte has been activated, it undergoes an affinity maturation process, through mutation and selection of the genes that produce immunoglobulins. This process is called somatic hypermutation and requires an interaction between an antigen-activated B lymphocyte and an activated $T_{FH}$ cell (although, as we shall see, other types of T-helper cells can also induce this). The completion of this process results in the selection of B lymphocytes that have now generated high affinity receptors for the epitope to which they initially bound with low affinity. Now, a single antibody molecule, an isolated IgG, with only two antigen-binding sites, will be able to bind to it strongly enough. This allows the activation of different phagocytic cells, particularly those with Fc receptors for IgGs, and the implementation of more effective mechanisms to fight viruses and bacteria.

## 3.- Slowing down the initial impulse

Up to this point we have talked about the mechanisms that are set in motion during the activation of the different cells of the immune system to deal with infectious microorganisms. We have also discussed that the action of these cells, necessary as it is to eliminate infections through the inflammatory response, also causes collateral damage in the form of tissue degradation and exerts effects on blood circulation and on the distribution of body fluids through tissues. For this reason, it is necessary to slow down the immune response once the threat has been overcome. Otherwise, the cells would remain continuously activated and producing cytokines and chemokines that would lead to increased activation of the immune system unnecessarily. This would increase collateral damage and eventually also cause autoimmunity, that is, cause some T lymphocytes to become active against one or another of our own antigens.

For this reason, there are also mechanisms for inactivating the immune response when it is not needed. The understanding of these mechanisms has allowed the development of new anti-tumor strategies that are proving to be very effective and, therefore, I believe it is very interesting to know about them.

The inhibition mechanisms act mainly on the activated T cells, since these are the ones that generate and secrete the molecules that activate the different processes that the immune response puts in place against the different classes of microorganisms that threaten us. On the other hand, to survive, B cells, in addition to stimulating cytokines, need the continuous presence of the antigen that they recognize. This antigen, when interacting with the B-cell receptor activates it and thus provides the intracellular biochemical signals allowing the survival of the B cell, which would otherwise die from apoptosis. Once the antigen has been eliminated, most B cells that are actively producing antibody cannot receive these surviving signals and die of apoptosis. However, not all of them die, since some lymphocytes, in the process of their activation, become memory B cells, which remain ready to be activated quickly in case they detect their antigen again in the future. As we will see later, memory cells are the reason for the effectiveness of vaccines.

In addition to the disappearance of the antigen they recognize, caused by its elimination thanks in part to its activity in secreting antibodies against it, the B cells must also stop receiving the signals provided by the stimulating cytokines, which are produced by helper T cells. To do this, the latter must be actively inhibited, since once they are activated, not all of them cease to be or die when the antigen disappears.

Remember that for proper activation, T cells need to receive three signals from antigen-presenting cells. One of them is received through the T-cell receptor, another one is received through the cytokines secreted by the antigen-presenting cells and the last one is received through the CD28 receptor expressed by the T cell, which is activated by the co-stimulatory molecules B7-1 and B7-2 expressed by the antigen-presenting cells. This last signal is the most important to favor the survival of the activated T lymphocytes. They also provide survival signals to B cells through the expression of the CD40L ligand for the CD40 receptor expressed on B cells. If T cells do not receive the co-stimulatory signal via CD28, they do not express adequate CD40L levels, so they cannot help B cells produce antibodies, nor can they help macrophages kill extracellular bacteria.

For this reason, inhibition of the co-stimulatory signal sent by the CD28 receptor is one of the most important mechanisms used by the immune system to curb excessive T lymphocyte activity. The T lymphocytes themselves, after their initial activation, begin to express inhibitory receptors on the membrane that result in the generation of inhibitory signals for cell growth and survival.

One of these inhibitory receptors is **CTLA-4**. This receptor, like CD28, interacts with the co-stimulatory molecules B7-1 and B7-2, but it does so with a greater avidity, so it is able to prevent CD28 from interacting with B7-1 and B7-2, since these molecules bind CTLA-4 with preference. In this way, the T cell cannot receive the signal through CD28 and decreases its activation level.

T cells when activated do not initially express CTLA-4 on the membrane. This molecule is found in intracellular vesicles. However, some time after activation of the T lymphocyte by an antigen-presenting cell, CTLA-4 is transported to the membrane, where it interferes with the binding of B7-1 and B7-2 to CD28.

The amount of CTLA-4 located on the membrane depends on the level of activation received by the T cell. The higher the initial activation level, the more CTLA-4 ends up on the membrane and therefore the higher the inhibition level received. In this way, the T cell can regulate the intensity of the activation signal and keep it at appropriate levels. This is important because it has been shown that laboratory mice that have had their *CTLA-4* gene removed die very young from uncontrolled activation of their T cells.

How does this mechanism of automatic regulation of T lymphocyte activation work? Like everything else in the cell, the mechanism depends on chemical reactions and interactions. Chemistry may not be a fairly appreciated branch of science, but I can assure you that without chemistry there would be no life, because life, intelligence and also human feelings and emotions would not exist without chemical interactions.

We are not going to enter here into the complex and fascinating world of the regulation of the chemical interactions taking place inside a cell in response to an external molecular signal received by a receptor, but we can mention that a very widespread form of molecular activation with which cells modulate chemical interactions and thus finally achieve that the external signal reaches the cell nucleus and modifies the expression of genes, is **the phosphorylation of proteins**. The phosphorylation of proteins consists of the addition of so-called **phosphate groups** in specific amino acids of these (serine, threonine or tyrosine, for those who want to know). The phosphate groups consist of a central phosphorus atom surrounded by four oxygen atoms ($PO_4$) and have electrical charge. The amino acids to which the phosphate groups are added have no electrical charge, i.e. they are neutral. When the phosphate group is added to them, however, the amino acids change their mass and their charge, acquiring two negative charges that they did not have before. The presence now of these two negative charges allows new electrostatic interactions with other amino acids, either from the same or different proteins. In either case, the phosphorylated protein is thus activated and is now capable of exerting a function that it could not exert before.

Phosphorylation is therefore a chemical modification that works like a switch, "turning on" proteins that were previously "off".

Like all chemical reactions that take place inside cells, phosphorylation is an enzyme-catalyzed reaction. The particular class of enzymes that catalyze phosphorylation reactions are called kinases. These are further divided into two main classes: **tyrosine kinases**, which add phosphate groups to the amino acid tyrosine, and **serine/threonine kinases**, which add phosphate groups to the amino acids serine and threonine. T- and B-cell receptors carry these types of enzymes bound to their intracellular regions and when they interact with an antigen this interaction results in the activation of the bound kinases. These kinases phosphorylate, that is, they add phosphate groups, to specific proteins, which eventually leads to the activation of one or several transcription factors that travel to the nucleus and modify the expression of certain genes, for example, cytokine genes, among others.

The CD28 receptor of T cells is one of the kinase targets. When it binds to B7-1 or B7-2, CD28 is phosphorylated, which in turn leads to the activation of another kinase that is essential for signal transmission to the cell nucleus. Similarly, the CTLA-4 molecule is a target for kinases activated by the TCR and CD28 receptors and is also phosphorylated by them. When CTLA-4 is phosphorylated, this chemical modification allows CTLA-4 to reach the membrane. This does not happen if CTLA-4 is not phosphorylated. Therefore, activation of the receptors, which in turn activate the kinases, results in the phosphorylation of CTLA-4, which will then be located on the membrane, where it will inhibit CD28 activity.

CD28 inhibition will result in less phosphorylation of CTLA-4. This lower phosphorylation will eventually lead to less CTLA-4 on the membrane and thus to greater activation of CD28. Thus, the mechanism of CD28 activation itself leads to its own inhibition, and inhibition of this receptor by CTLA-4 also leads to its own activation. This allows for a self-regulation of T lymphocyte activation levels, which reach a state of equilibrium and remain there. This state leads to a lower production of the cytokine IL-2 in cells that have been activated for a longer time than in newly activated T cells, which do not yet express CTLA-4 on their membrane and are, therefore, more sensitive to activating signals. As we know, IL-2 is a fundamental factor for the clonal expansion of T cells, so the action of CTLA-4 is essential to limit the excessive proliferation of T cells after their activation, which can lead to the development of autoimmunity and leukemias.

Other receptors that inhibit T lymphocytes are more active than CTLA-4. The latter seems only to compete with CD28 for interaction with B7-1 and B7-2. However, other inhibitory receptors have their own ligands and are active in slowing down the signal of the activating TCR and CD28 receptors. The way they do this is by activating one or more enzymes that counter kinase activity. These enzymes are called **phosphatases** and their activity leads to the removal of the phosphate groups from the proteins to which the kinases have added them. **Dephosphorylation**, as this removal is called, returns the activated enzymes to their resting state.

One of the most important inhibitory receptors that act through the activation of phosphatases and which is expressed by T cells is **PD-1** (Programmed Death 1, although despite this name its activity is not directly related to apoptosis). PD-1 receptor expression is substantially increased on antigen-activated T cells and B cells, but this receptor is not expressed by macrophages or dendritic cells. Like CD28, PD-1 has two ligands, called **PD-L1** and **PD-L2** (Programmed Death Ligand 1 and 2), which are proteins alike B7-1 and B7-2 and are therefore said to belong to the same protein family. PD-L1 is expressed permanently by many cells in the body, including macrophages, dendritic cells, and B and T cells. PD-L2 is expressed only by dendritic cells and macrophages and both types of cells increase their expression significantly in response to inflammatory cytokines such as IFN-γ and IL-4.

The above data indicate that while the PD-1 receptor expression increases on activated T and B cells, at least one of their ligands, PD-L2 also increases its expression on antigen-presenting cells when they receive cytokines produced by activated lymphocytes. From this we can conclude that the very activation of lymphocytes and the production of cytokines by them induces their own deactivation through increased expression of PD-L2 in dendritic cells. This inhibition is due to the activation of phosphatases by the PD-1 receptor.

Still another inhibitory receptor that works by phosphatase activation is **BTLA** (B and T lymphocyte attenuator). This inhibitory receptor is expressed on both activated T and B lymphocytes, and it is also expressed on some cells of the innate immune system. The ligand for this receptor does not belong to the same protein family as that of the other inhibitory receptors, and when it is activated it acts on the molecular mechanisms that lead to the activation of the transcription factor NF-κB, a very important factor not only for the correct activation of innate immune system cells, but also of B and T lymphocytes when these find their antigen.

Other inhibitory intracellular proteins do not act on the signals sent through antigen receptors or co-receptors, but their action is focused on inhibiting the signals sent by some activating cytokines. These proteins are called **SOCS** (Suppressors Of Cytokine Signaling). These proteins contain domains capable of binding to certain regions of tyrosine phosphorylated proteins and sequestering them, thus preventing them from exerting their activity. The production of SOCS proteins is stimulated by the cytokines themselves, whose activity they must inhibit, so that once the signal sent by the receptors of these cytokines has reached the cell nucleus, in addition to inducing the synthesis of effector genes, it also induces the synthesis of the SOCS proteins, which allow the action of the cytokines on the cells to last only for a limited time.

Finally, the most decisive mechanism for slowing down the immune response once the infection has been overcome is the elimination of the cells that carry it out, especially T and B lymphocytes. This occurs by inducing apoptosis, which can be provoked both by the absence of stimulating signals sent by antigens and by signals inducing this cell death process sent to activated lymphocytes by other cells. Among the latter is the signal sent by the **Fas receptor** activated by its **FasL ligand (section 2.6)**. Throughout the immune response, Fas expression is stimulated in activated lymphocytes and makes them more susceptible to die by apoptosis through interaction with cells expressing FasL. Mutations in the Fas receptor that prevent it to send proapoptotic signals cause the so-called **autoimmune lymphoproliferative syndrome**, characterized by attacks of the immune system to our own tissues and organs.

We see, therefore, that inhibitory receptors intervene in multiple molecular aspects related to the activation of immune cells or their elimination. Together, these receptors act to stop an excessive activation of lymphocytes or cells of the innate immune system, which would cause various diseases and pathological situations.

### 3.1.- Macrophage polarization

An aspect also related to the regulation of the inflammatory response is the phenomenon of macrophage polarization. This phenomenon consists of the fact that, depending on the external signals they can receive, these cells are able to modulate their properties and perform different functions. These signals lead to the generation of two main types of macrophages, called **M1 macrophages** and **M2 macrophages**, which can be converted into each other depending on the circumstances.

M1 macrophages are the classic macrophages, i.e. those that we have previously found fighting at the source of infection. They are activated in the course of infection in response to the detection of bacterial or viral components. They are activated by the signals emitted by TLRs when these components are detected **(section 2.5.1)**. Similarly, cytokines produced by $T_H1$ or NK cells **(section 7.4)**, particularly IFN-$\gamma$, TNF-$\alpha$ or **granulocyte-macrophage colony stimulating factor (GM-CSF)**, which we have not discussed so far, can also stimulate macrophage activation and turn it into an M1-type macrophage.

M1 macrophages are specialized in the fight against invasive pathogens, although it has been proven that they can also participate in the fight against tumors. They are cells with a high phagocytic capacity and, in addition, can act as antigen-presenting cells, traveling to the lymph nodes with the lymph, as we have also seen. They produce cytokines that have a pro-inflammatory activity, that is, they stimulate the fight against infections. These cytokines include TNF-$\alpha$, IL-1, IL-6, IL-12, and IL-23, which also promote the activation of adaptive immunity by stimulating the expression of MHC-2 and co-stimulatory molecules B7-1 and B7-2, in addition to the CD40 receptor. The cytokines secreted by M1 macrophages also help to recruit monocytes from the blood to the sites of infection, where they will be converted into macrophages that will help in the fight against infection. Similarly, M1 macrophages secrete chemokines that help monocytes and $T_H1$ lymphocytes reach the sites of infection.

M1 macrophages are the type of macrophages responsible for the phagocytosis of the bacteria and the triggering of the respiratory burst **(section 2.2)**. They can also produce other pro-inflammatory substances, such as certain prostaglandins, derived from arachidonic acid, a fatty acid synthesized from another essential fatty acid, linoleic acid, which must be acquired through the diet. Prostaglandins are the main inducers of fever, which is also part of a mechanism that enhances adaptive immunity.

M2 macrophages, unlike M1 macrophages, are anti-inflammatory, that is, they will act to stop the inflammatory response when it is no longer necessary because the infection has been overcome. They are generated in response, among other signals, to IL-4 and IL-13 cytokines, produced by certain CD4 T cells, and by some CD8 T cells, as well as by the M2 macrophages themselves. The main function of these macrophages is the resolution of the inflammatory response. To this end, they produce anti-inflammatory cytokines, such as **IL-10** and **TGF-β**, and other substances also derived from arachidonic acid that are anti-inflammatory. M2 macrophages do not appear to act as antigen-presenting cells, although they still possess high phagocytic capacity. They use this capacity mainly to eliminate remains of dead apoptotic cells and neutrophils. At the same time, they possess regenerative capabilities for the tissues that may have been damaged during the inflammatory response.

It has been proven that M1 macrophages can become M2 and vice versa, depending on the signals that these cells receive from other cells. This repolarization provides functional versatility that allows macrophages to adapt to possible changes in the course of infection.

We can conclude here our first walk through the mechanisms of the immune system. Now we can go into some surprising depths of it. Let's start by exploring in more detail how the immune system differentiates between self and non-self, and even differentiates between healthy self and ill self, a capability without which we would all be dead.

## 4.- B-LYMPHOCYTE RECEPTOR DIVERSITY

We have explained that each B lymphocyte has a receptor on its surface capable to binding to some molecule or part of a molecule that is or could be found in the external world. We also said that some of these receptors may even bind to substances that do not yet exist, but which could be synthesized in the future by some laboratory. We are now going to explain how each B lymphocyte can generate a receptor that fits and binds one molecule or another, and how the billions of B lymphocytes in an organism, all together, have enough receptors to detect and bind to any substance, known or unknown.

To understand this mysterious matter, it is useful to dwell for a moment on concepts that may be familiar to the reader, but which must be remembered. As we have said, for a lymphocyte to generate an antibody, its receptor must detect an epitope and must physically attach itself to it. Physical binding requires the formation of chemical bonds which, in this case, normally depend on a difference in electrical charges (whether this difference is temporary or permanent) between the surface of the receptor and the surface of the epitope. The receptor and epitope surfaces must be complementary to each other, i.e. they must be like two pieces of a puzzle that fit together. In addition to being complementary in shape, the electrical charges of one of the surfaces must be opposite to those of the other surface, although this electrical charge difference is often only generated when both surfaces are sufficiently close to each other (and, for those who want to know, the so-called van der Waals bonding forces, among others, come into play). In any case, the difference in electrical charges is what creates a enough bonding force between the epitope and the B-cell receptor.

For the above reason, the ability of a B-lymphocyte receptor to form bonds with a given epitope depends on the physicochemical properties of the amino acids that form that receptor. The receptors are proteins, and, like all proteins, they are formed by the concatenated binding of amino acids. There are about twenty amino acids in nature, which show a great diversity of chemical properties. Indeed, there are those with positive charge, with negative charge, without charge, with affinity to water, that repel water, etc. These last properties in relation to water are very important, since all the interactions between the molecules of life take place in an aqueous medium, which strongly influences whether a certain interaction can happen or not.

If we understand the above, we will understand that the ability of a receptor to interact with any substance will depend on the type of amino acids it shows on the surface of the binding site, which must be complementary, as we said, to that of the epitope it can detect. Therefore, the diversity of epitopes that can be detected will depend on the combinations of amino acids that can be presented at the area of the receptor surface intended to interact with the epitope. For example, if this surface shows negatively charged amino acids, it will only be able to interact with epitopes from a complementary but positively charged surface. If the surface of the receptor shows negatively and positively charged amino acids at different locations, it will interact with a negatively and positively charged epitope at locations complementary to those on the surface of that receptor. The figure below illustrates the idea of complementarity between epitopes and antigen-binding sites of light and heavy antibody chains.

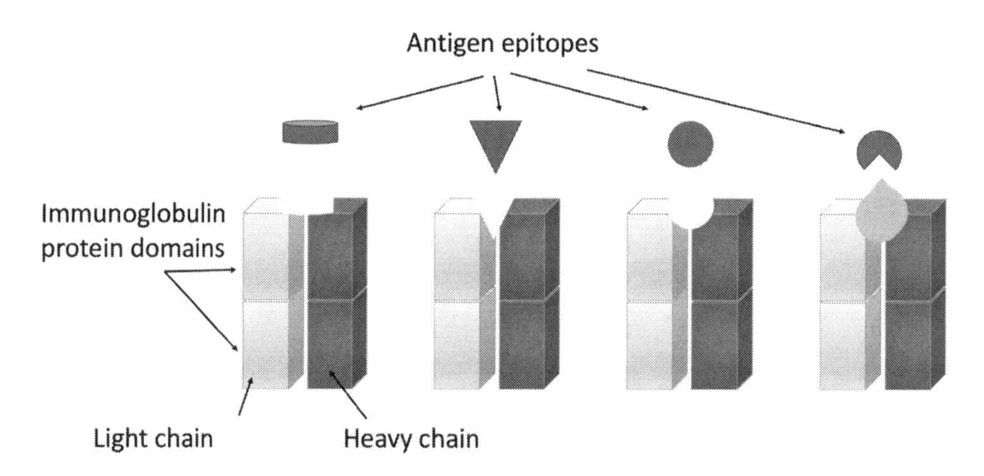


***Antibody Fab fragments adopt different shapes on their surface adapted to the epitopes of each antigen***

From the above, we can deduce that, to generate protein receptors against any substance we can imagine, we only have to devise a system that allows the placement of the greatest possible diversity of amino acids in an area of the external surface of the receptor, while maintaining a common structure for those receptors. This common structure is important because it is what allows them to interact with the proteins of the cellular machinery, which in this case are all identical in the different B lymphocytes, to send a signal to the cell nucleus when the receptor has detected its epitope. The idea is, therefore, to generate a receptor protein that, while maintaining areas of common amino acids, with no or minimal variation among them, contains areas on its surface where the variation of amino acids is maximum. The areas of minimal variation serve to maintain a common structure adapted to the function that the receptor (and later the antibody when produced) is to perform. The zones of maximum variation are those that serve to have billions of receptors that can each interact with a given molecule out of the billions of unknown substances that exist in nature. By generating billions of receptors with these characteristics, we can be sure that some of them will detect a substance characteristic of any virus, bacterium, fungus, etc., and may serve to neutralize the threat.

Since each protein needs a gene to be produced, the mystery was how it was possible for B lymphocytes to have so many different genes. The answer to this mystery took decades of research, but, in short, the answer revealed that each B lymphocyte, during its maturation from a stem cell to an adult B cell, builds a different gene for the heavy chain and another for the light chain of its receptor and antibody (which are produced by the same gene) from the selection of dozens of pieces of DNA, which, combined together in a unique way in each lymphocyte, produce mature genes with different information in specific areas. These genes will be able to direct the production of the millions of different receptors with the diversity of amino acids we are talking about precisely in the region of the receptor destined for interaction with the epitope, while respecting an identical general structure for the receptors and the antibodies. Let's see how the precursor genes of the immunoglobulin light and heavy chains are organized and what happens so that each lymphocyte generates a different version of mature genes for each of these chains.

## 4.1.- Recombination of antibody genes

The discovery of the structure of the precursor genes for the antibodies heavy and light chains had to wait for the development of the technologies of molecular biology and DNA sequencing. These technologies have led to the discovery of this structure, which is fascinating and provides a good example of the generation of diversity by applying simple rules to a set of elements that combine with each other.

The precursor genes of the B-cell receptor heavy and light chains are made up of a series of diverse DNA regions that must be cut and joined together, i.e. recombined, to generate a mature gene. In the case of the heavy chain gene (located on human chromosome number 14, in a place called the heavy chain gene ***locus***), these regions are called **V** (for variability), **D** (for diversity), **J** (for joining) and **C** (for constant). The union of a **V** region, with a **D**, with a **J** and, finally, with a **C** region, is necessary for the generation of a heavy chain mature gene. There are five main classes of **C** regions for heavy chains, which are named with the Greek letters mu (μ), delta (δ), gamma (γ) –of which there are four subclasses–, epsilon (ε) and alpha (α) –of which there are two subclasses–. This makes a total of **nine C regions for the heavy chain**. However, the light chain precursor genes, of which there are two, named with the Greek letters kappa (κ), –located on human chromosome number 2–, and lambda (λ) –located on human chromosome number 22–, contain only regions **V**, **J** and **C**, but no **D** regions. The **C** regions of the light chains are different from those of the heavy chains and do not affect the final class of antibody produced. In this case, a mature light chain gene is generated by the binding of a **V** region to a **J** region and to a **C** region (either from the κ *locus*, or λ *locus*). The following figure explains the process of this union of **V**, **J** and **C** regions, of this genetic recombination, as it is called in the language of science, for the case of the light chain.

***The combination of V, J and C regions forms the mature light chain gene***

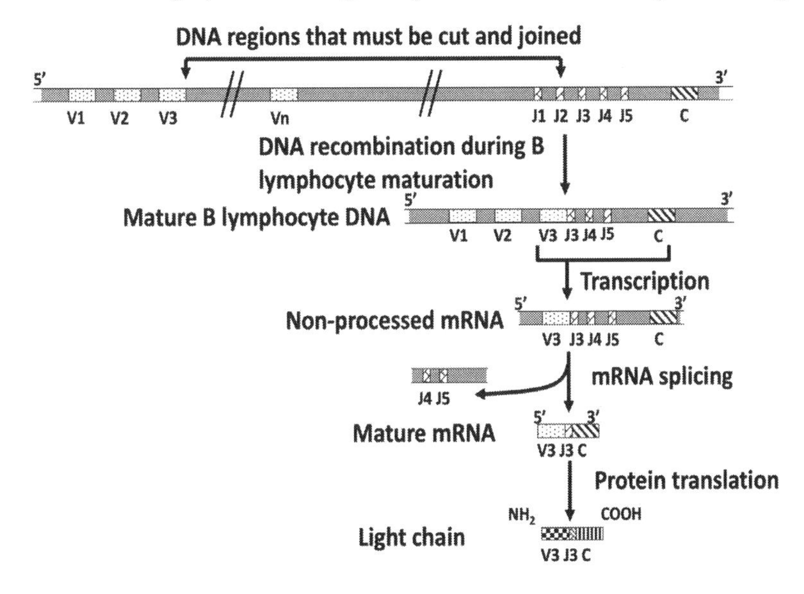


It is clear that, since the **V**, **D** and **J** regions are recombined by randomly choosing one of each class in each lymphocyte and putting them together, depending on the amount of **V**, **D** and **J** regions available in the genome, a greater or lesser amount of different mature genes for antibodies will be obtained and, therefore, a greater diversity of antibodies. However, it is not desirable to have more **V**, **D** and **J** regions than are necessary to generate enough antibody diversity to cover all possible molecular structures to be recognized. Having too many regions, in addition to the cost in terms of DNA synthesis, which costs energy, also increases the complexity of managing that precursor DNA to generate mature genes with only one **V**, **D** and **J** regions for the heavy chain or only one **V** and **J** regions for the light chain of the antibodies. Therefore, we should ask ourselves how many **V**, **D** and **J** regions have been generated throughout the evolution of our genome to achieve an adequate diversity of antibodies against external threats.

Interestingly, data collected from sequencing the genomes of hundreds of individuals have revealed that we do not all have the same number of **V** fragments in our genome. The number of **J** regions and **C** regions for the light chain class genes also shows a slight variability between individuals. However, we all have 23 **D** regions, 6 **J** regions and 9 **C** regions in the heavy chain precursor genes. If you are curious about the maximum and minimum amount of **V**, **D**, and **J** regions that an individual may have, consult the following table (data extracted from Janeway's Immunobiology, 9th Edition).

| Number of DNA fragments in the human immunoglobulin genes | | | |
|---|---|---|---|
| Fragment | Light chains | | Heavy chain |
| | κ | λ | H |
| Variable (V) | 34-38 | 29-33 | 38-46 |
| Diversity (D) | 0 | 0 | 23 |
| Joining (J) | 5 | 4-5 | 6 |
| Constant (C) | 1 | 4-5 | 9 |

How does each B cell choose which regions to join? How is it guaranteed that B cells that have joined regions differently will be produced so that each can produce an antibody against a different epitope?

To understand this, we must consider several factors. The first is that in the bone marrow there are stem cell precursors of B cells that are constantly dividing and generating a multitude of daughter cells. These daughter cells go through a series of maturation stages in which they will generate the mature genes for the heavy and light antibody chains. Each daughter cell, from which millions and millions are produced, has the opportunity to randomly select **V**, **D** and **J** regions to generate a particular heavy chain, as well as the possibility of randomly selecting **V** and **J** regions to generate a mature light chain gene. In this case, moreover, the cell will generate a mature light chain gene of only the κ type or only the λ type, that is, it does not generate two mature genes for two light chains, but only one. In the end, the mature B cell has one mature heavy chain gene and one mature light chain gene and can use these genes to generate a receptor ready to detect a certain epitope. This epitope could perhaps be found in some microorganism that might try to infect us if we encounter it at a moment of our lives. This will depend on the vicissitudes of everyone's life. To understand why, here's an example: suppose a B lymphocyte in your body has a receptor capable of detecting an epitope present in a microorganism found only on the island of Papua New Guinea. Under these conditions, unless you decide to travel to Papua New Guinea and during that trip you are affected by that microorganism, your B lymphocyte will never be activated because it will never find the epitope capable of doing so. However, the immune system continuously generates billions of "just-in-case" B lymphocytes that, hopefully, will never be of any use.

How then does each B cell derived from a precursor stem cell generate a particular mature gene for the light chain and another specific mature gene for the heavy chain? Well, it does so in the simplest way: by random combination of one **V** region with one **D** region and one **J** region, **always in that order**, in the case of the heavy chain, and the random combination of one **V** region with one **J** region for the light chain. In other words, no two **V** regions ever join, no two D fragments ever join (although there are some exceptions to this latter rule) and no two **J** fragments ever join. Each **V**, **D** and **J** fragment is flanked by regions with a specific nucleotide sequence (the "letters" of the DNA) that serve to indicate to the recombinant enzymes, i.e. the enzymes that cut and paste the DNA, at which locations and in what order they should make the cuts and joins. The reason that the **V**, **D** and **J** fragments, in the case of the heavy chain, or **V** and **J**, in the case of the light chain, must be joined in that order is that, if the joining were to take place in a different order, the information that would end up in the mature gene would not make sense and would not generate the correct protein.

To better understand what happens when the random combination, albeit in a defined order, of several DNA fragments occurs, we can make use of an analogy using words. Suppose we must construct a multitude of simple sentences with the subject, verb and complement structure: for example, "the dog is brown". To build the sentences we have 60 nouns,

2 verbs and 23 complements. We can randomly choose any of the 60 nouns. Likewise, we can choose any of the two verbs and any of the 23 complements. However, if we want the sentence to make sense, we have to choose a noun first, then a verb, and finally a complement. Although we can understand phrases such as "brown is the dog", or "good is the thing", cells are much dumber than we are and can only understand phrases in the subject-verb-complement order and nothing else, i.e. they can only understand the information generated in the DNA when a **V** fragment is combined with a **D** fragment and a **J** fragment in the case of the immunoglobulin heavy chains, or a **V** with a **J** fragment in the case of light chains.

Therefore, although we can randomly choose any of the **V**, **D** or **J** regions, we can only join them in that order if we want the "sentence" they generate to make sense. How does the cellular machinery choose one or another fragment? Well, since the choice is made practically at random (although there may be some factors that favor some recombinations better than others), whether one or another fragment is chosen depends, above all, on whether the enzymes that catalyze DNA cutting and joining reactions bind to one fragment rather than another. Once these enzymes, called **recombination activation gene (RAG)** enzymes (of which the most important are called **RAG-1** and **RAG-2**) have bound to DNA, they first produce a cut in the DNA just behind one of the **D** regions and another cut just in front of one of the **J** regions, in the case of the heavy chain gene. The DNA separating both regions is removed and the two cut ends are now joined together. In this way a randomly generated **DJ** fragment has now been formed. The RAG enzymes will then bind behind one of the randomly selected **V** regions and cut the DNA at that location; the enzymes will also bind just before the **DJ** fragment and cut the double strand of DNA there as well. The cut DNA fragment that separated the **V** and **DJ** regions is removed, and the ends of the DNA now behind the **V** region and before the **DJ** region are joined. We have thus formed a **VDJ** region, which will be part of the mature heavy chain gene. This gene, in addition to the **VDJ** region, needs the binding of a **Cμ** region to generate a heavy chain of a B-cell receptor that will at the same time generate IgM antibodies if the B cell is activated by an epitope to which the receptor can bind. In summary, a mature gene for a heavy chain of the B-cell receptor (and of the IgM antibody) has been formed by random gene recombination of regions **V**, **D** and **J** and their binding (which this time is not random) to a **C** region. The figure below shows this process for the heavy chain of antibodies.

***The combination of V, D, J y C regions forms the mature gene for the immunoglobulin heavy chain***

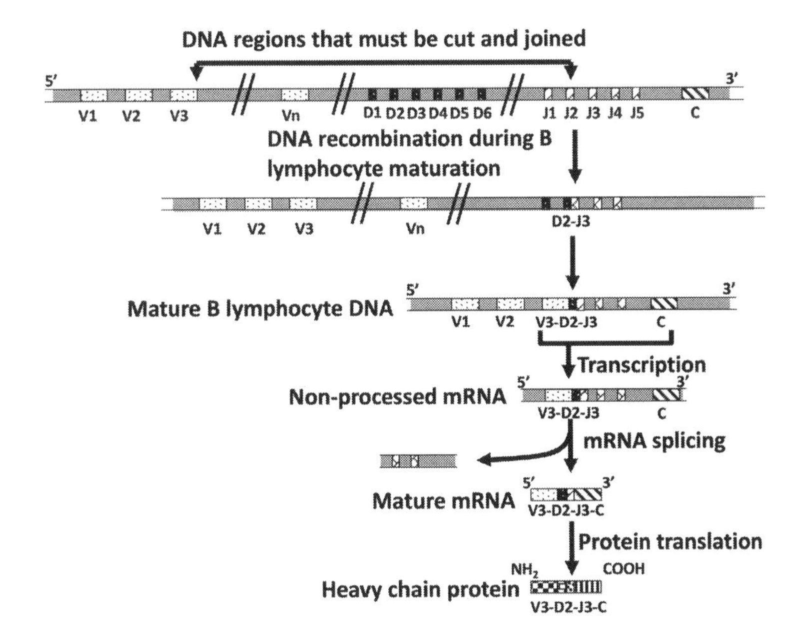


## 4.2.- Allelic and isotypic exclusions

So far, we have only briefly seen the process of recombination to generate the heavy and light chains of immunoglobulins. However, this is not the only molecular process involved in the correct generation of antibody chains. We will briefly explain here two additional processes that are necessary for a B cell to become a full-fledged lymphocyte.

As we know, each lymphocyte must possess one receptor for one epitope, and only one. If it possessed two different receptors, the lymphocyte would produce two kinds of antibodies when it was activated, but one of the antibodies produced would be useless. The reason for this is that to be activated, it would be enough for the lymphocyte to find one or the other of the epitopes that it could detect with any of its two receptors. The activation of a lymphocyte with two receptors by only one of the epitopes would lead to the generation of a clone of activated B cells that would produce and secrete large quantities of the two antibodies, because the genes of the receptor being formed in the same way, they could not be differentiated when activated. Under the above conditions, half of the energy used for the B cells to produce antibodies would be wasted.

Perhaps here we ask ourselves the question of whether it is possible for a single B cell to generate two different receptors. Well, yes, it is possible. Let's remember that each cell has two copies of every gene, that is, two alleles of every gene, one inherited from the father and one from the mother. Therefore, each immature lymphocyte possesses two copies of the genes for the heavy chain and four copies for the genes for the light chains, because, let us remember, these were of two types, κ and λ, and each of them also possesses two alleles, one maternal and one paternal.

Throughout evolution, those individuals unable to maximize the use of the always scarce resources for survival have been eliminated. In other words, the organisms able to ensure that each B cell produces antibodies with maximum efficiency in the face of a specific threat and does not waste resources are those that have been able to survive to this day.

To guarantee this maximum efficiency, evolution has generated two fascinating mechanisms that operate in B cells, and we will see that one of them also operates in T cells. The first of these mechanisms is **allelic exclusion**, which operates in both heavy and light chain genes. The second is **isotypic exclusion**, which operates only in light-chain genes.

Allelic exclusion is a molecular mechanism by which one of the two alleles of the light or heavy chains of immunoglobulin genes is silenced as soon as the cell has successfully recombined the allele on the other chromosome. The process of recombination begins with the joining of one **D** and one **J** regions of the heavy chain genes on one of the two chromosomes. Once these regions are joined, a **V** region is attached to them. This process begins first in one of the chromosomes, believed to be the first to be generated in the process of cell division that takes place in the generation of daughter cells from the precursors in the bone marrow. Thus, one of the two chromosomes is ahead of the other in the process of recombination, so that one chromosome recombines before the other. If this recombination is successful on the first allele (success is not guaranteed in all cells that initiate the recombination process), a complete heavy chain is produced. This heavy chain is transported to the cell membrane from where it is already able to send a biochemical signal into the cell nucleus. For this signal to be produced, the heavy chain must first bind to two other proteins that the developing B cell produces with the sole mission of detecting when the cell has successfully produced a heavy chain. These proteins are called **λ5** and **VpreB**. The two together form a kind of surrogate light chain (remember that the B lymphocyte has not yet rearranged the light chain gene, which will happen only after checking that the rearrangement of the heavy chain genes has worked correctly). The binding of the newly formed heavy chain to this surrogate light chain forms a kind of temporary receptor (called a **pre-B receptor**) on the cell surface. This receptor does not detect an antigen but, thanks to the surrogate light chains, is able to interact with itself. This interaction of the pre-B receptor with itself provides information indicating to the cell that it has correctly formed a heavy chain. The information is transmitted from the membrane to the nucleus, where mechanisms are set in motion to prevent recombination of the other heavy chain allele that might be occurring on the other chromosome. This leads to the generation of B cells that have only one of the heavy chain alleles rearranged in a functional way.

Obviously, if the first allele has not been successfully rearranged, i.e. if the **VDJ** recombination has happened with errors that prevent the formation of a correct heavy chain, the pre-B receptor is not formed, and the cell continues with the rearrangement of the second heavy chain allele. In this case, two things can happen: either the rearrangement is successful, and a correct heavy chain is produced at this second opportunity, or this rearrangement also fails, and the cell cannot generate the heavy chain. If the former happens, the pre-B receptor formed will send the signal to the nucleus. In this case, the signal will be unnecessary to silence the other allele, but again, it will serve the function of informing the cell that it has correctly rearranged a heavy chain allele, even if it was at the second opportunity. This signal is fundamental to allow the cell to remain alive, since, if at this moment of its life the cell does not manage to recombine one of the two heavy chain alleles, the cell "knows" that its mission in life will not be able to be fulfilled. The cell will be useless and, in that case, will commit suicide through the process of apoptosis. All these processes lead to the generation of B cells that have only one of the heavy chain alleles rearranged in a functional way. The following figure represents the process of allele exclusion in B cells.

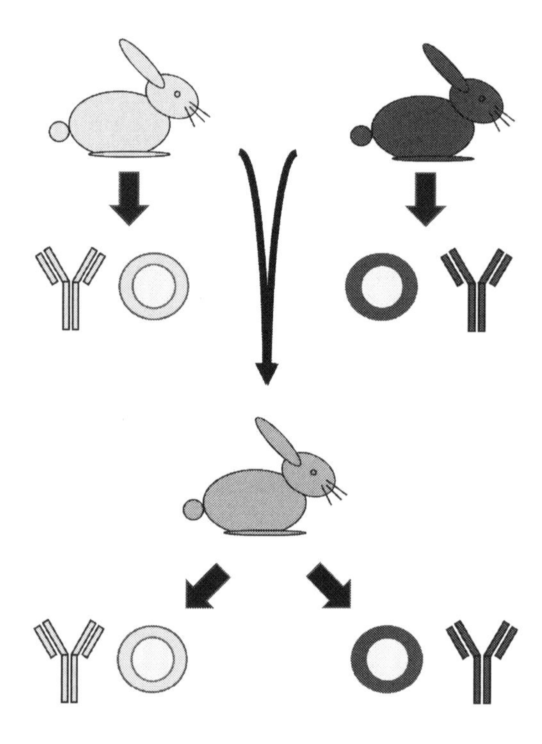

The figure shows the crossing of two laboratory rabbits which are homozygotes for two different varieties of the heavy chain genes of immunoglobulins, represented here by two different shades of grey, light and dark. The hybrid offspring generated by the cross (intermediate shade) has a copy of the maternal allele and a copy of the paternal allele. If the generated B cells did not suffer from the allelic exclusion described above, each would have receptors with heavy chains generated from either of the two alleles, the one inherited from the father or the one inherited from the mother. However, analysis of mature B cells from hybrid rabbits indicates that the mature B cells from this animal have heavy chain receptors that come from only one of the alleles; the other has therefore been excluded. This exclusion of one or the other of the alleles occurs randomly, according to which of the two chromosomes begins the process of recombination first.

The mechanism for ensuring that B-lymphocyte receptors have only a light chain is somewhat more complicated. Remember that these receptors, and the antibodies that are derived from them, may possess light chains of one or other of the two isotypes, κ or λ. However, receptors and antibodies only possess one type of light chain, which is, moreover, identical in both arms of the molecule, i.e. it is always κ or always λ. In other words, there are no immunoglobulins that have a κ chain on one arm of the Y, and a λ chain on the other. This means that, in addition to the developing B cell having to ensure that light chain genes are rearranged on a single chromosome to the exclusion of the other, it is also necessary for the cell to set in motion a process to ensure that one of the two light chain isotypes is excluded in the formation of the mature antibody molecule. This is called **isotypic exclusion**.

To ensure this exclusion, the cell has developed mechanisms for sequential control of the light-chain gene recombination. Once the cell has checked that it has correctly rearranged the heavy chain, the rearrangement of the κ light chain alleles, in one or another of the chromosomes where they are located, begins. If this process is successful on the first allele, the rearrangement stops. If it is not successful, the κ chain allele on the other chromosome is rearranged. Again, if this process is successful, the rearrangement is stopped. If, on the other hand, it fails again, one λ chain allele in one of the chromosomes is rearranged. If it succeeds, the recombination of the other allele is stopped; if it fails, the other allele is recombined. If one rearrangement process is successful, eventually the B cell will have achieved a mature light chain gene and will generate a mature B-cell receptor. If the process is unsuccessful, the cell will not be able to generate a functional B-cell receptor, so it will not be able to receive the **survival signals** that this receptor sends, and the cell will not be able to prevent the apoptosis process from being triggered and will die.

Thus, all these processes ensure that during the development of B cells, each one produces a single B-cell receptor directed against a single epitope or, at most, against epitopes that are closely related chemically and spatially. If this were not the case, these processes also guarantee that no useless cells will be generated for the organism, cells that would be costly to maintain but which, nevertheless, could not contribute to its defense. The process of apoptosis in the absence of the survival signals sent by the pre-B or mature B-cell receptor guarantees that these useless cells are eliminated. Consequently, all B lymphocytes that are generated have functional B-cell receptors that may be able to find a single epitope on the surface of a foreign microorganism and contribute to the defense against it.

This leads us to remember that, as we said before, the strategy that could allow the generation of receptor proteins and antibodies against practically all substances needed a structure that would allow focusing all amino-acid chemical diversity on certain areas of the receptor surface. The structure, at the same time, should have a common shape, since all antibodies and receptors are very similar molecules. It is now time to address the analysis of this structure and to explain how the **V(D)J** recombination (the parenthesis indicates that light chains do not recombine D fragments) allows the production of B-cell receptor proteins with a common structure that, however, concentrate a huge diversity right in the areas of interaction with the epitopes.

### 4.3.- The immunoglobulin domain

As mentioned above, the B-lymphocyte receptor and the antibodies are proteins that present varied amino acids in an external zone destined for interaction with the epitope. These amino acids will be able to form chemical bonds with some of the atoms that form the epitopes in antigen molecules. However, the amino acids that form any protein will not only establish links with external molecules, but most of them will establish links with amino acids of the protein itself. These links between many of the amino acids of a protein make it fold in space and acquire a three-dimensional structure that is normally adequate for the protein to perform its function in collaboration with other proteins or molecules. The three-dimensional structure that a protein acquires depends on the sequence of amino acids that it possesses, since according to this sequence the different chemical properties of each amino acid follow each other, which, in the aqueous medium in which the proteins are found, determines how these interact with other neighboring or distant amino acids and also determines which amino acids of a particular protein do not interact with any other of the protein itself. These amino acids that do not interact with those of the protein itself are those designed to interact with other molecules, including other proteins or DNA.

In the case of immunoglobulins, it has been shown that the amino acid chains that form them, due to their precise sequence, fold in space into three-dimensional structures that have been called the **immunoglobulin domain**. Protein domains are areas of proteins that fold in space independently from other areas of the same chain of amino acids that form the complete protein. Proteins can possess several domains, each named after a specific protein, characterized by its three-dimensional shape. In our case, immunoglobulins only possess domains of this name. One reason for this is that the structure of the immunoglobulin domain holds the secret to the enormous diversity of antibodies and, at the same time, to their common structure. Let's see how this domain is folded in space.

The key to the functionality of the immunoglobulin domain is that the amino acids that form it fold into a structure that has been called the **β–sheet.** This "sheet" is formed by folding the chain of amino acids forming turns every few amino acids, like that shown in the figure below.

***Structure of the β- sheet protein folding***

The arrows indicate the direction of the amino acid chain, which always has a starting point and an end point, of course. Likewise, the length of the arrows indicates the areas of the amino acid chain that establish links with each other, which fix the folding in this way. In the case of the figure, an arrow in one direction establishes interactions with neighboring arrows, which go in the opposite direction. As can be seen, we can thus divide the amino acid chain folded in this way into two classes of zones: the zones that establish interactions between them, which correspond to the arrows, and the zones that do not establish interactions with other zones of the amino acid chain, which correspond to the turns (represented by thin fragments that join the arrows). Since these turns are not necessary for the maintenance of the structure, which only depends on the interactions between the amino acids located in the arrows, any amino acid can be placed in them, whatever their chemical properties. This is not possible in the areas of the protein that correspond to the arrows, because if we change the amino acids in them we change the chemical properties, which could prevent the correct formation of the links that maintain the arrow zones joined to each other forming the sheet. Since the turns are oriented towards the exterior, these will be in charge of establishing the interactions with external molecules.

The immunoglobulin domain follows these design rules. It is formed by two β-sheets that are placed one over the other forming a kind of sandwich, as shown in the following figure, which represents an immunoglobulin domain of one of the variable regions of the immunoglobulins.

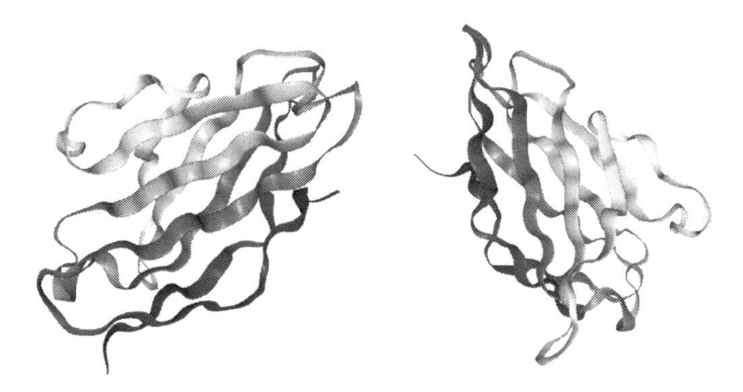

***Two views of the same immunoglobulin domain variable region. The ribbons represent the amino acid chain and how it is folded in space.***

The folding of the immunoglobulin domain always places three of the turns of the sheet outwards. These are the ones that will interact with the epitopes present on the surface of the antigens. Whether they interact with an epitope or not depends on the class of amino acids located at the turns, but since they can appear in all possible combinations, and since they are also flexible and can adapt to a large number of different shapes, the potential number of epitopes with which they can establish interactions is virtually unlimited.

We should expect that, if we compare the DNA sequences of the already recombined immunoglobulin mature genes, or the heavy and light chain amino acids that are derived from the sequence of these mature genes, we should expect that some areas would be much more different than others. These zones should also coincide with the turns of the immunoglobulin domains, since it is in these turns that the greatest diversity of amino acids is found. Indeed, when this analysis was performed, this is what was observed. Heavy chain **VDJ** regions or light chain **VJ** regions possess three areas of high sequence diversity of both DNA "letters" and amino acids in the case of the proteins. These regions are called **hypervariable regions**, since they show enormous variability in the corresponding nucleotides and amino acids.

For each mature immunoglobulin chain protein, we have three hypervariable regions that, indeed, coincide when folded in space with the three turns of the immunoglobulin domain. These regions are called **CDR1**, **CDR2**, and **CDR3** (Complementarity Determinant Regions. We speak of complementarity because these regions are complementary to the form and chemical properties of the epitope). Two of these hypervariable zones are encoded in the DNA of the **V** regions present in the genome. The third hypervariable zone, **CDR3,** however, is encoded in the **V-D**-joining zone (in the case of the heavy chain gene) or the **V-J**-joining zone (in the case of the light chain). Since changes in the DNA sequence of the joining zone occur during recombination, with enzymes randomly adding or removing some "letters", this third hypervariable region is the most hypervariable of all and, therefore, the one most responsible for the enormous diversity of antigen receptors and the antibodies derived from them and their ability to bind to virtually any molecule of the external world.

A final consideration should be mentioned here. It is clear that if one B-cell receptor is randomly generated in each one of the billions of B lymphocytes produced in the bone marrow, since there is no restriction on the amount of substances that these receptors as a whole can detect, this means that some of the mature B cells may possess receptors that detect our own molecules. If this happens, those B lymphocytes are eliminated or inactivated by different mechanisms that we are not going to talk about here. Let's stay with the idea that through random gene recombination during the development of B lymphocytes, a repertoire of billions of B lymphocytes is generated with receptors capable of detecting virtually any molecule present in the external world, but which are not normally capable of detecting or reacting against our own molecules. We will see that a similar, though more sophisticated, process takes place in the thymus for the selection of the T-lymphocyte repertoire, which we will discuss below.

## 5.- Identity masks

In the previous section, we have seen how B lymphocytes are capable of generating immunoglobulins against virtually any molecule found in the world, and even against molecules that, although not found, will be found in the future, as a new drug yet to be invented or discovered. However, for the immune system to function properly, in addition to being able to identify what is foreign to the organism, it must also be able to identify what is self in a healthy self in order to distinguish it from what is self in a sick or unhealthy state, for example, when a cell in the organism has been infected by a virus, or when a cell has become malignant. It is, therefore, necessary for the organism to generate its own signs of identity and for the immune system to learn to distinguish them in states of cellular health and disease.

These marks of identity are fundamental, since multicellular organisms are based on the collaboration of cells of identical genome. The maintenance of an identical genome among all the cells of an organism has been so important throughout evolution that, although the cells have specialized and acquired specific functions and in no case have all the genes of their genome working at the same time, they still maintain all their genes. It would be much cheaper (though perhaps more complex) for a given organism if the cells of the liver, for example, lost the genes they didn't need and the cells of all the other organs did the same. In this way, growth from the fertilized zygote to a mature organism could be achieved at a much lower metabolic cost, since, if from the early stages of cell differentiation and growth these unnecessary genes were lost, a significant amount of energy and matter would be saved in embryonic and postnatal development. Undoubtedly, this saving could provide an evolutionary advantage to the organisms that had adopted it. However, none of the known multicellular organisms have adopted such a strategy. All cells have the same genome and the genes are silenced or activated depending on the type of mature cell involved and the function it is to perform.

Why is this so?

Well, I don't think the answer is completely known, but it is believed that in the case of cells, collaboration between equals is so important that multicellular organisms could not have been generated otherwise. And equality between cells is measured by equality in their genomes. Probably, moreover, the mechanisms by which genes are acquired or lost from a given genome throughout evolution have not allowed for a different situation from the one we have today, which implies that practically all the cells of an organism have the same genome (with some exceptions such as precisely mature lymphocytes, that have recombined genes and lost part of their genome, and also the red blood cells, which have lost the cell nucleus and therefore their entire genome), and that the specialization distinguishing some cells from others is done, as we have mentioned, by activating or silencing the genes.

In any case, to ensure collaboration between equals, it is necessary for the cells to manifest their identity and confirm that, in fact, they belong to the same multicellular organism derived from a single primordial cell, that is, that they all belong to the same family. As in the case of human societies, where it is usually the police who control identity, control of cell identity is also carried out by a "cellular police", which in this case consists of T lymphocytes.

The system developed for the communication of cell identity to the T lymphocyte "police" is fascinating and one of the most complex and difficult to understand in the immune system, since it is responsible for eliminating any cell that does not prove that it belongs to the organism (that is why graft rejection occurs, for example) or that does not prove that it has not been subverted and become a dangerous cell for the others (such as a cancer cell or a cell infected by a virus). How do cells indicate their identity and goodwill to the cells of the immune system and thus avoid being killed? Well, they do so thanks to a set of molecules that we have already seen as part of the response to infection: the molecules of the **major histocompatibility complex**, **MHC**, which are present on the surface of all cells and are detected by T-lymphocyte receptors. The MHC molecules are the repositories of information about the identity of each cell and the T lymphocytes are responsible for detecting this information and acting accordingly, allowing life or causing the death of the cells, depending on their interpretation of this identity. We will see, however, that the MHC molecules do not actually indicate the identity of the cells of the organism to those of the immune system, but that the latter normally ignore normal cells and can only detect those cases in which the cell has changed its personality, has "radicalized" itself, and is beginning to be dangerous to others. It is this change of identity that, under normal conditions, the cells of the immune system can detect, and the way in which this is achieved is really a marvel of Nature. Let's try to explain it with an amusing analogy.

## 5.1.- The Facegood tribe

To begin to understand how this system works, let's dive into an allegory and travel to the heart of an unfortunate tribe of blind and deaf people who, nevertheless, need to know at all times who are members of the tribe and who are not. Let us remember that cells are also blind and deaf, and they can only touch each other or "smell" the substances they emit, such as cytokines or chemokines.

Let's call this tribe the Facegood tribe. A terrible illness, of unknown cause, has led to all members of the tribe losing their senses of sight and hearing. They possess only the senses of smell and, above all, touch to identify each other.

It is vital for the members of the tribe to identify those who belong to it. The tribe lives surrounded by other enemy tribes. With patience and much collaboration, the Facegood tribe members were able to build a wall that protects their domain. The wall has a gate through which, at night, some detachments go out to try to collect forest fruits, firewood and other materials necessary for the survival of the tribe. On their return, they obviously need to get through the gate and this is the moment when infiltrators from other tribes, who are obviously neither seen nor heard, can take advantage of the opportunity to easily penetrate the wall through the gate at the same time as the detachment enters.

These enemy infiltrations caused much misery and pain and seriously endangered the survival of the tribe. The infiltrators were dedicated to taking advantage of the tribe's scarce resources and living at their expense, which became untenable if the number of infiltrators became too high. The survival of the Facegood tribe was threatened. That is how things were until the greatest genius the Facegood tribe ever had was born. This chosen one of the gods, named Vedejay, had the most brilliant idea ever to inhabit a human mind: to make killer masks.

A killer mask is a special kind of mask that this genius devised. It is a mask made of fine silicone (modern tribes have access to all kinds of materials) that fits so perfectly on a human face that it leaves no room for air to enter, so the unfortunate one who has a face where the mask fits perfectly will die of suffocation in a few minutes. Only people with faces in which the masks don't fit well would leave a little room for air to enter and would not suffocate.

Obviously, the problem with this approach was how to make killer masks that would only kill those who were NOT members of the Facegood tribe. Once the masks were made that way, the idea was to establish a mask police that would continually test masks at random on people within the tribe's compound. That way, sooner or later they would meet an intruder, if there was one. As long as no mask fit a face, the people tested would not die, and that would be because that face would belong to a member of the tribe and not to a foreign invader. However, if any killing mask fit someone's face, it would be because that person was an invader who had infiltrated the group and was not a member of the tribe. In that case the killer mask would suffocate him or her.

The idea was good, because it meant that all members of the tribe would always be controlled with respect to their identity and thus, sooner or later, parasite infiltrators who sought to take advantage of them would be identified and eliminated. The problem, however, was that, in order to make killer masks, Vedejay only had the faces of the tribe members as a mold, and not the faces of the members of other tribes. How could he make masks that would fit unknown faces that he did not have access to, but at the same time prevent them from fitting the known faces of his tribe members?

At the moment he was considering this, Vedejay got his genial idea. He realizes that instead of making masks using only one known face as a mold, what he has to do is generate masks by randomly mixing the features of his tribe members' faces. In this way, masks would be generated that would not fit well on any of their faces, always leaving a gap where air could pass through. These masks, as long as they contain all the components of a human face, nose, mouth, cheekbones, etc..., will fit sooner or later in some unknown face. Thus, he made multiple silicon replicas of the noses, ears, cheekbones, etc. of all the members of the Facegood tribe. In addition, he randomly modified each component of the faces, thus generating a great diversity of ears, noses, etc., generally different from those belonging to the members of his tribe. By means of the random combination of the components thus generated, Vedejay manufactured innumerable masks that could potentially fit any face, although normally they could not fit perfectly in the faces of any member of the tribe, unless, by bad luck, all the components of the face of some unfortunate one had been combined, so that a killer mask had been generated for him that sooner or later would end up killing him.

However, at this point, Vedejay had two problems. The first was to check that the masks he had made had turned out well and that they were indeed masks that fit human faces, not masks so defective that they would not fit any. If there were many such imperfect masks, they would greatly diminish the effectiveness of his system, so he had to remove them from the collection of masks generated. The second problem was to eliminate those masks that, since they had been generated at random, might fit perfectly on the faces of some members of his tribe, since they could kill them.

To solve these problems, Vedejay had another great idea. He called the official sculptor of the tribe and asked him to make plaster faces identical to that of each tribal member. On these faces he could now try out his masks without danger of suffocating anyone.

Once the plaster faces were ready, Vedejay began to test his masks on them. Vedejay first made a selection that he called **positive selection**, that is, he selected all those masks that fit more or less well on the plaster faces and eliminated those that did not fit well at all. Again, if the masks didn't fit the faces of his tribe members to some degree, it was because they were not suitable masks; they were too defective, and they probably wouldn't fit any other face either. As a result, in a first selection, Vedejay was left with only masks that fitted at least to an acceptable degree with the plastered faces.

But this wasn't enough. Now it was necessary to make a selection that Vedejay called **negative selection**. This selection consisted of eliminating from the population of masks that had been previously selected those that fit too well on the faces of the members of his tribe. Vedejay tested the masks again on the faces and this time he did it in a more detailed way to check the degree in which each one of the masks fit in each one of the plaster faces. Any mask that fit too well was removed.

Vedejay was thus left with a collection of masks of which he was sure, firstly, that they were well-made masks, that is, that they had the property of fitting perfectly into some human face, and, secondly, that there was no mask that fit perfectly on the faces of the members of his tribe. With these masks, the Facegood members would prove the identity of their faces at every opportunity.

The procedure was a success. Little by little, he managed to eliminate all those who were infiltrating the tribe by posing as members of the detachments. There was always some mask that fit well enough to suffocate a stranger, and in general, they never asphyxiated the tribesmen. The only exceptions were, unfortunately, when one of the members of a night detachment hit his nose hard and returned with a swollen one. The unfortunate man had the misfortune that a mask fit his deformed face perfectly and smothered him. Vedejay instructed the members of his tribe that, in the event of an accident, they were not to return unless they were certain that his face had not been damaged. If it had, they were to wait until healed before returning to the village. Otherwise it could mean death for them. Despite these difficulties, the Vedejay system proved to be extremely effective in protecting the members of the Facegood tribe, the only tribe of blind and deaf people that has been able to survive to this day.

## 5.2.- Molecular "faces" and "masks"

With these ideas in mind about the Facegood tribe, we now have better intellectual tools to understand how the immune system examines the identity of the body's cells, and not only the identity in a state of cellular health, but also the change in identity that can occur when the cell is forced to produce foreign proteins, having been infected or having mutated into a malignant cell. Similarly, the immune system examines the changes in cell identity that can occur when cells pick up foreign proteins produced by infectious microorganisms, which can also change the molecular identity of the cell that has picked them up. Which molecules make up the "faces" and which ones make up the "masks" so that the blind and deaf cells of the immune system can examine the identity of self to distinguish it from the non-self?

Let's start with what constitutes the "faces". These are made up by the molecules of the type 1 and type 2 major histocompatibility complex, which we have seen before **(section 2.5.4.1)** in the context of fighting infection. We will analyze how these molecules are formed and expressed on the cell membrane and why there are two classes of them, each of which represents a different type of "face" (as the faces of white people represent a type of face that differs from the type of face of yellow people, for example), for which it is necessary to generate two types of "masks", also different.

### 5.2.1.- The major histocompatibility complex

The major histocompatibility complex (MHC) is the set of genes responsible for generating the proteins that will function as the "faces" of cell identity. As we have said, there are two types of "molecular faces" made up by the MHC-1 and MHC-2 molecules.

Each mature MHC molecule of any type is made up of a combination of three components. Two of them are fixed for each type of person and cell, and one of them is variable.

#### 5.2.1.1.- The "faces" of class 1 MHC molecules

The two fixed components of this type of MHC are two protein chains, each derived from a gene. In the case of MHC-1 molecules, the first chain is called the **α chain** and the second chain is called **β2-microglobulin**. As in the case of immunoglobulins, with their heavy and light chains, we have here a new molecule formed by the union of two different chains. We see then that this is a recurring theme in the immune system and many immune system molecules are formed by the union of two different proteins, and sometimes by three or more. In the case of MHC-1 molecules, both chains are linked together and provide stability to each other. However, the α chain is perhaps the most important, because it is responsible for binding the third component of the complex. This component is no other than a peptide derived from the degradation of some of the proteins produced by the cell from the functioning of its genes or genes of external parasites, such as virus. We already have seen that, in the case of a viral infection, these peptides can be derived from the proteins of the virus that have been produced by the protein synthesis cellular machinery, but in the case that the cell has not been infected, its MHC-1 molecules show, nevertheless, on the cellular surface, peptides coming from the degradation of the self-proteins that the cell produces. These peptides, linked to α chains, which are in turn also linked to β2-microglobulin, reveal on the cell surface the information about the state of the cell. The following figure represents an MHC-1 molecule with the peptide attached to its binding groove.

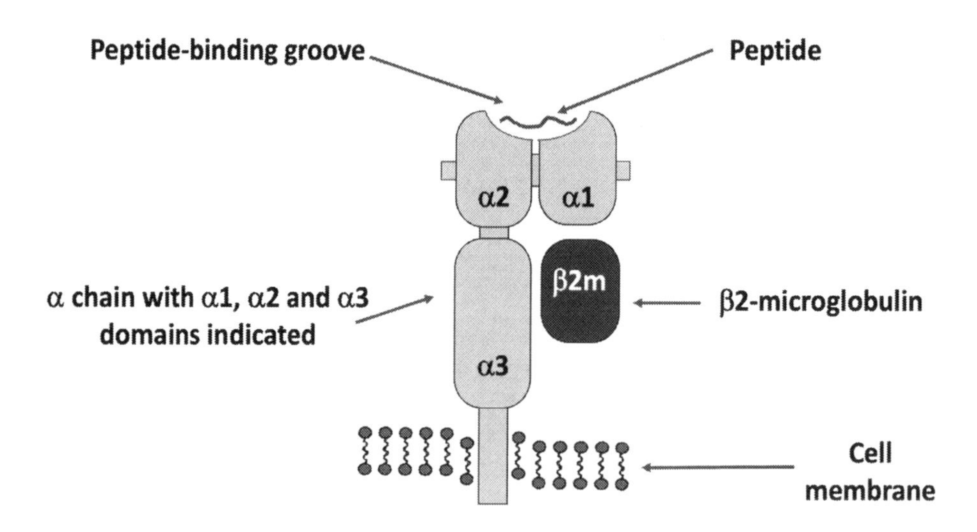


***Representation of the MHC class 1 molecular structure***

#### 5.2.2.2- The proteasome

Where do the peptides that bind to MHC-1 molecules come from and how do they combine with it inside the cell? Well, peptides come from all kinds of cellular proteins that, for whatever reason, have stopped working properly (usually because they have been totally or partially denatured, or because they have been fragmented) or have never worked, because they have been incorrectly synthesized, and must be recycled for the generation of new proteins. This is done by digesting the amino acids chains that form the proteins to release the constituent amino acids, which can then be used for the synthesis of new proteins. This digestion is carried out by enzymes known by the generic name of **proteases**, which in this case are different from the metalloproteases released by the cells of the immune system at the site of infection **(section 2.5)**. The degradation of proteins is carried out in stages, and the first stage consists of the fragmentation of the proteins into short peptides so that they can then be hit by other enzymes to release the individual amino acids.

The degradation of useless proteins is so important that, as we have said, cells have an organelle specialized in this function: the so-called **proteasome**. This organelle is also involved in controlling the concentration of undamaged proteins, so that they are always at optimal concentrations in the cell.

The proteasome is like a complex factory dedicated to the recycling of proteins. We will briefly describe how it is formed so that we can once again glimpse yet another of the marvelous mechanisms that cells harbor and the equally marvelous research work that has made it possible to identify these and other cellular structures and their mode of action.

The proteasome is made up of two parts: a nucleus and two regulatory heads, one of which is important to allow the entry of proteins into the nucleus, where degradation occurs, while the other is located where the peptides generated after degradation will exit. The nucleus is formed by the union of twenty-eight protein subunits, fourteen subunits called $\alpha$ and fourteen called $\beta$.

The core is like a kind of sandwich with two slices of bread and two slices of ham. Seven of the $\beta$ subunits form a ring, and the other seven form another ring. Both rings are placed together, side by side, forming the inner core. Now, seven α subunits form another ring and are placed on one of the rings of the inner core. The remaining seven α subunits form another ring and are placed on the other ring of the inner core. In this way, the core of the proteasome is formed, consisting of four rings, two β rings (the slices of ham), and two α (the slices of bread), one on either side of the β rings.

Thus, the nucleus of the proteasome is a kind of hollow seven-sided cylinder. It is within this kind of cylinder formed by the fourteen inner β subunits and the fourteen outer α subunits that the degradation of proteins into peptides of only seven or eight amino acids in length takes place. It is the inner subunits that act as the protease enzymes responsible for the degradation of proteins into peptides. Let us remember that a peptide is simply a small fragment of protein.

Let us dwell on the result of this degradation. Whether young or old, proteins are generally made up of hundreds or even thousands of amino acids. In the proteasome, these proteins will be cut into small pieces (peptides) from only seven or eight of these amino acids. The cuts are made almost randomly, so the same protein of hundreds of amino acids in length will not necessarily produce identical peptides. In addition, some of the peptides generated in a first cut can be linked to others to generate somewhat longer peptides that no longer correspond to a particular amino acid sequence in the protein being degraded. In this way, a large number of different peptides are produced from the degradation of the molecules of the same class of protein, and the same happens with other types of proteins, of course. From the set of all proteins produced by the cells, millions of different peptides are produced. These peptides are going to be coupled with MHC-1 molecules that will be transported, each one with its charge of a peptide, to the cell surface. There, they are going to constitute a kind of fingerprint, of molecular barcode of the cell identity, that is to say, the molecular identity of the cell, since not all types of cells produce the same proteins, nor do they produce them in the same precise amount. If this molecular fingerprint changes, even just a little, due to the presence of foreign peptides, for example, proteins from a virus that has infected the cell, the immune system will be able to detect this change and act accordingly.

Obviously, for proteins to be degraded in the proteasome, they must be captured and brought inside. In this task, one of the two heads, formed by ten protein subunits, plays an important role. This head is bound to the outer ring of α subunits on one of the sides of the core and acts as a kind of cover, as a mobile hood, which allows or not the access of the proteins to the proteasome entrance for their degradation. Nine other protein subunits form a similar structure attached to the other α ring at the other end of the proteasome, where it plays the role of facilitating the exit of the peptides generated inside. The following figure shows representations of the human proteasome.

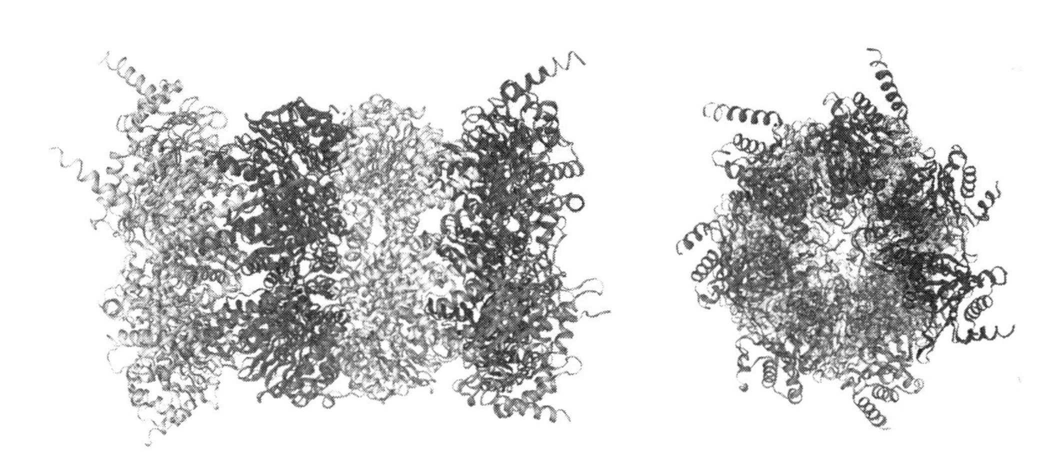

***Lateral and frontal view of the human proteasome central core***

#### 5.2.2.3.- Tagged for degradation

In general, for a protein to be introduced into the proteasome it must first be identified as defective or old by yet another cellular machinery specialized in this task. It is obvious that the cell must avoid unnecessarily degrading healthy and functional proteins, since it costs a great deal of energy to produce them. For this reason, the activity of the proteasome is finely regulated. The most important way in which this regulation is achieved is by tagging old proteins for degradation. Untagged proteins cannot enter the proteasome and cannot be degraded.

As everybody knows, tagging something means adding a mark to it. In the world we live in, a mark can be a painted or engraved sign on an object, but in the world of molecules, as proteins are, marks can only be made by other molecules. In the case of proteins that must be degraded, this mark is generated by the enzymatic bonding to them of a small protein called **ubiquitin**. The enzymes that bind ubiquitin to other proteins are called **ubiquitin ligases**. Ubiquitin is so called because when it was discovered, in 1975, it was found in virtually every cell in every organ. It was ubiquitous, and hence its name.

The addition of ubiquitin to an old or deteriorated protein is called **ubiquitination**. This addition happens thanks to a chemical reaction, which, like all chemical reactions in cells, is accelerated, catalyzed, by enzymes. In this case, the sequential action of three enzymes is necessary for the binding of a ubiquitin molecule to an old protein. Furthermore, these enzymes not only bind a ubiquitin molecule to the identified defective proteins, but they can also bind several molecules of this protein, one after the other, forming chains of ubiquitin linked to the defective protein and further ensuring its degradation by the proteasome.

Only the proteins that have been linked to ubiquitin are recognized by the part of the proteasome that functions as the gate, which, as we have said, regulates the entry of the proteins into the inner nucleus where their degradation takes place. This entry requires metabolic energy (adenosine triphosphate molecules, ATP, the universal currency of metabolic energy in all cells, are consumed). Ubiquitin functions as a key to the gate, unlocking it and allowing the protein to enter the nucleus of the proteasome, where it will be broken down into peptides. Non-ubiquitinated proteins do not possess this key and cannot enter the proteasome, which protects them from improper degradation.

Using again the analogy of faces and masks, MHC-1 molecules would constitute only one type of "face" (another type we will talk about later in more detail is MHC-2 molecules). However, these "faces" are incomplete. Making another analogy, we can say that MHC-1 molecules are "faces" without a "nose". The "nose" of each MHC-1 "face" will be formed by the peptides generated in the proteasome, which must be joined to the MHC-1 molecules to generate complete "faces". Thus, the MHC-1 "faces" present on the cell surface are similar in a given individual, and only differ because they possess millions of different "noses". These "noses" are formed by the universe of peptides generated in the proteasome, coming from the degradation of self or foreign proteins produced in the cell. The "noses", therefore, constitute a varied set of peptides that, together, contain the information about the identity of a cell. In the case that one or more of these peptides vary, this will be an indication that the cell has changed its identity, which can only happen if its health state has changed, either because it has been infected by a virus, or because it has been transformed into a tumor cell by mutations in some of its genes and proteins. These changes make the cells dangerous for the organism, so they must be detected and eliminated. Later we will talk about how the "noses" join MHC-1 "faces", a process that, if not working properly, can generate serious diseases and it is also a mechanism that many viruses interfere with to avoid detection.

#### 5.2.2.4- A "FACE" FOR EVERYONE

However, MHC-1 molecules are different in different individuals in a population. Just as each person has his or her particular face, each person has his or her given "molecular faces" in the form of MHC-1 molecules and the peptides attached to them. This is because the genes that produce MHC-1 proteins are very diverse, i.e. there are hundreds of different ones. Moreover, the human genome has not only one gene to generate the α chain of MHC-1 molecules, but three, with an allele of them in each chromosome, so that each person can generate six different MHC-1 α chains, that is, each cell displays six different types of MHC-1 "faces". In addition, there are hundreds of different alleles for each of the three MHC-1 genes. Therefore, it is very unlikely that two people, unless they are identical twin brothers, would have the same alleles to produce their "molecular faces", which must then be supplemented with each cell's own "molecular noses".

This is part of the reason why people, in general, are incompatible when it comes to donating or receiving organ grafts. Each person has its own "molecular faces", and these will be strange if introduced with a transplant into another person. In this case, it is certain that at least one of those strange "faces" will perfectly fit onto one of the "masks" that have been generated by the immune system of the graft recipient and that, as we have explained before, are designed so that they do not perfectly fit onto healthy self "faces", but onto self "faces" modified by the presence of different "noses", formed by peptides originated by the degradation of proteins of infectious microorganisms, for example. Let us remember that the selection of "masks"" (the receptors generated by T lymphocytes) so that they do not fit in one's own "faces" allows the existence of "masks" that fit well in slightly different "faces", whether this difference is due to the "nose" (the peptide) or to the rest of the "face" (the MHC-1 molecule). This perfect fit is what activates the action of the immune system, which will eventually reject the transplanted organ unless immunosuppressive drugs are used to prevent it and the immune system eventually tolerates it.

#### 5.2.2.5.- MHC-2 "FACES"

The "faces" made up of MHC-1 molecules are specialized in showing "noses" generated by proteins synthesized in the cell itself. Another type of "faces", those made up of MHC-2 molecules, are specialized in showing "noses" derived from proteins captured from outside the cell. How are these "faces" formed, and how are their "noses" generated?

As with MHC-1 molecules, MHC-2 molecules are made up of two protein chains called α and β. In this case, however, unlike MHC-1 molecules, both chains are involved in the binding of a peptide. The following figure represents the structure of an MHC-2 molecule.

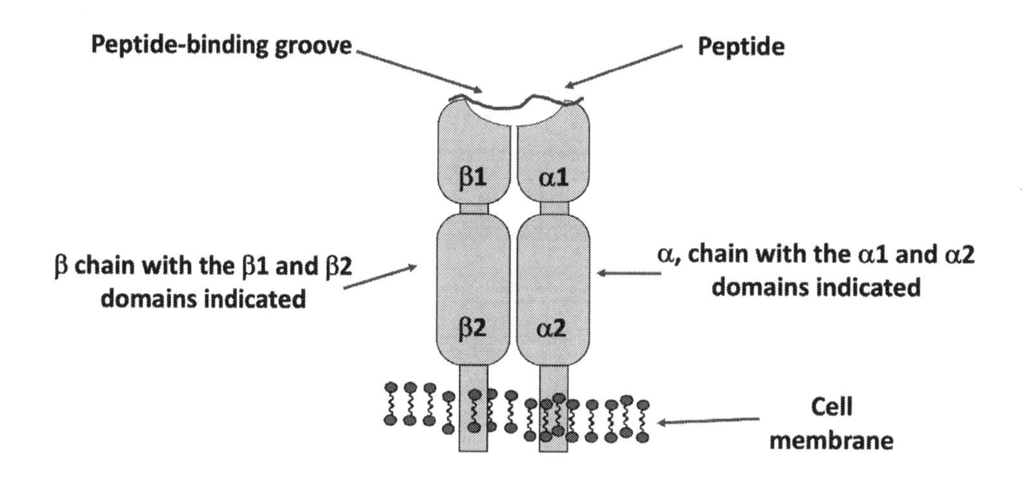


***Representation of the MHC-2 molecular structure***

Another difference from MHC-1 molecules is that both MHC-2 chains cross the cell membrane. However, the most important difference is in the origin of the peptides that bind to the MHC-2 molecules, which come from the digestion in the lysosomes of external proteins captured by the cell. This implies that only cells capable of incorporating proteins from the environment will be able to present peptides derived from them in the MHC-2 molecules. In fact, cells that cannot phagocytize or incorporate foreign proteins in any way do not generally have MHC-2 molecules on their surface, since they do not have the genes activated to produce the $\alpha$ and $\beta$ chains, since they do not need them. For this reason, under normal conditions, only professional antigen-presenting cells, that is, those specialized in the function of presenting the enemy to T lymphocytes, express the genes to produce MHC-2 proteins and present peptides in them. There are four types of these cells: dendritic cells, macrophages, B lymphocytes and eosinophils, which are cells involved in allergic reactions that we do not examine in this book.

The MHC-2 molecules expressed in these cells may contain peptides derived from the body's self-molecules captured by them from the extracellular medium and fluids. These self-peptides are generally ignored by T cells, since they have been selected to fit only weakly into our own "faces". Only when the antigen-presenting cells capture molecules from foreign microorganisms, they can present "noses" that manage to turn the MHC-2 into foreign "faces", which will lead to the activation of T lymphocytes and defense mechanisms. These "noses" are also different from the "noses" presented by the MHC-1 molecules. It should be remembered that the latter are generated in the proteasome, while those presented by MHC-2 are generated in the lysosomes. This results in very different peptides, particularly in their length. Peptides bound to MHC-1 are shorter and more homogeneous, with only 8 to 10 amino acids, and fit perfectly into the cleft in which they must be coupled to be presented. The peptides presented by MHC-2 are more diverse in length and in some cases the longer peptides overflow the cleft and have their ends outside it.

### 5.2.2.6.- MHC GENES ARE POLYMORPHIC AND POLYGENIC

MHC genes are a magnificent example of how the evolution of animal genes works when they are under great selection pressure, i.e. when individuals which do not possess certain variants of those genes or the right amount of them are rapidly and relentlessly eliminated from the population and therefore cannot pass on their genes to the next generations.

This selection pressure is very strong on the MHC genes. The reason is simple to understand: these genes are required for the very important mission of placing peptides derived from external pathogenic microorganisms on the cell surface to present them to T cells, without which adaptive immunity cannot function. Adaptive immunity, however, generates a great selection pressure on the microorganisms it attacks and, in general, eliminates. Obviously, under these conditions, those microorganisms that better can avoid the presentation of their antigens may have a better chance of survival and will transmit their genes more often. We will see later **(section 7)** that microorganisms have developed throughout their evolution numerous and ingenious mechanisms to avoid the action of the immune system. Several of them act on the mechanism of peptide presentation.

In this situation, an ongoing evolutionary war takes place between microorganisms and the animal organisms they parasitize and sicken. This war has had and will continue to have important consequences for the evolution of the MHC genes and explains two important properties of these: **polymorphism** and **polygeny**. Let's start by explaining the first one.

As we know, the prefix *poly* means "many", and the suffix *morphism* means "form". Therefore, the term *polymorphism* means "many forms". This means that the human population, and that of most vertebrates, possess multiple variants of MHC-1 and MHC-2 genes. This is important. The multiple forms of these genes are found in the population as a whole, but each individual contains only a few of these forms in its genome, not all the variants present in the population.

It is helpful to explain what differentiates one form of MHC from another. It is not difficult to understand that, if proteins are made up of amino acids, the differences between proteins will be mainly due to differences in the amino acids (although there may also be differences due to chemical modifications of these once incorporated into the protein chain). In the case of MHC proteins, however, the differences between their amino acids do not occur at random places, but precisely at the places responsible for binding self or foreign peptides, that is, in the amino acids that form the peptide binding cleft, the part of the "face" where the "nose" should fit. This implies that each form of MHC molecules has amino acids with slightly different chemical properties (electrical charge, affinity for water, etc.) in those places, which affects the nature of the peptides they can bind. Obviously, the peptides preferentially bound in the cleft will be those that possess amino acids with chemical properties complementary to those of the amino acids located at the cleft of the MHC proteins. Since each MHC protein possesses different amino acids with different properties in its peptide-binding groove, the peptides bound by one or the other of these forms will not be the same. This, in turn, implies that each individual in the population, by having its own forms of the MHC genes, will not present to the T cells exactly the same peptides derived from a certain microorganism. However, these differences, under normal conditions, have no influence on the ability of the immune system to control infections.

Now let's talk about polygeny. This term refers to the existence in each individual's genome of many MHC genes. This means that each person or animal possesses not just one, but several genes from which the MHC-1 and MHC-2 proteins are generated. In the case of humans, each of us has three genes to synthesize the MHC-1 α chains (β2-microglobulin is not polymorphic). These genes are called ***HLA-A***, ***HLA-B*** and ***HLA-C***. The letters **HLA** are derived from *Human Leukocyte Antigen*, the name by which human MHC proteins are also known. Since we inherit one variant of each gene from our father and another from our mother, we each have two variants, i.e. two alleles of each of these genes. Consequently, we have two *HLA-A*, two *HLA-B* and two *HLA-C* alleles. Each of these alleles is expressed in a co-dominant manner, i.e. both the paternal and maternal alleles are expressed in equal amounts and generate the same amount of protein. Therefore, each of us has in each of the antigen-presenting cells six slightly different types of MHC-1 molecules, each capable of capturing and presenting a population of peptides that are also different. This increases the number of peptides that can be presented by MHC-1 molecules, which decreases the probability that some microorganism could avoid the presentation of at least one of the peptides derived from its proteins by the antigen-presenting cells and thus escape the action of the immune system.

The case of MHC-2 genes is even more complicated, since MHC-2 proteins are formed by two chains, α and β, and each of them needs to be produced by a different allele. As in the case of MHC-1 molecules, each individual has three different genes for each chain. These genes are called ***HLA-DPA1***, ***HLA-DQA1*** and ***HLK-DRA***, which produce the α chain, and ***HLA-DPB1***, ***HLA-DQB1*** and ***HLA-DRB***, which produce the β chain. In the latter case, there is more than one *HLA-DRB* gene. All people have the *HLA-DRB1* gene, but some also have additional genes for this chain, called *HLA-DRB*3, *4* and *5*. This means that, in this case, since each person has two chromosomes, up to 12 different β chains (six from genes inherited from the father and six from the mother) can be combined with six different α chains (three from alleles inherited from the mother and three from the father). Thus, we see that the combinatorial diversity in the case of MHC-2 genes, as they are formed by two chains involved in peptide binding and presentation, is much greater than the diversity of MHC-1 genes.

The matter becomes even more complicated if we consider the enormous polymorphism of both *HLA-1* and *HLA*-2 genes. These genes are the most polymorphic known, that is, the ones with the most variants, the most alleles. Some have thousands of variants, while others have only a few dozen. This enormous polymorphism practically guarantees that there are very few people in the world who have exactly the same *HLA-1* or *HLA-2* alleles. In addition, the different variants of the *HLA* genes accumulate the differences in the amino acids involved in the union and presentation of peptides. This is a defensive advantage for the population as a whole, since if infected by a microorganism and suffering an epidemic, such as the COVID-19 pandemic that appeared at the end of 2019, caused by the SARS-CoV-2 virus, some of the individuals will be more effective than others in presenting peptides derived from the infectious microorganism and activating the immune system somewhat better. These individuals will be, in part for that reason, those who will survive the epidemic and the population will not become extinct. This illustrates one of the dangers of cloning. A cloned population, from the point of view of the immune system, would be like a single person. If this immune system were vulnerable to any microorganism, the entire clone population would be wiped out if an epidemic broke out. The polymorphism of the MHC genes guarantees that the population will have a great diversity in its ability to present peptides and activate the immune system, which guarantees that a single microorganism will not be able to kill it all. This also illustrates the risk suffered by species with few individuals, i.e. with low biodiversity, which are on the verge of extinction. The danger comes not only because the small number of individuals limits their reproductive capacity, but also because their genetic diversity, particularly that of the immune system, may not be sufficient to guarantee their future survival in the face of attack by microorganisms.

Finally, it is also necessary to talk about the concept of **MHC haplotype**. The prefix *haplo* means "half", so haplotype means "half the type". Since the MHC genes are all organized on the same chromosome, chromosome 6, we inherit half of them from the mother and half from the father. Each person thus has two different haplotypes, formed by the combination of the *HLA-1* and *HLA-2* alleles found on the chromosomes. Obviously, there are thousands and thousands of different haplotypes in the population, as each haplotype may differ only in one of the variants of one of the MHC alleles. This leads to most individuals being heterozygous for the MHC haplotypes. The only exceptions to this rule are identical twins and siblings who have inherited by chance the same chromosomes 6 from their parents. This situation makes it very difficult, if not impossible, to find fully compatible organ donors for people who need a graft, if they do not have a compatible sibling.

At this point, we will briefly describe the interesting process by which "masks" are generated and selected, that is, T cells are formed, each with a different receptor capable of interacting, although not very strongly, with our molecular "faces", that is, with our own MHC molecules of both types. However, in the case of some T cells with the appropriate "mask", this interaction will increase greatly in intensity when self "faces" have been modified by foreign peptides. To do this, we will rely on what we have already seen for the generation of B cell receptors and antibodies, since the process is very similar and involves practically identical molecular mechanisms.

## 5.3.- Generation and selection of molecular "masks"

To understand how each T lymphocyte in our body ends up possessing a unique receptor that will detect an MHC-1 or MHC-2 molecule, we need to first stop and analyze the molecular structure of this receptor. Remember that the function of any receptor is to detect a molecule outside the cell and transmit the information to the cell nucleus, where the cell turns on or off the genes needed to cope with what the information the receptor has detected advises the cell to do. In the case of the T-cell receptor, this implies that it must possess in its structure two different types of components. First, it must possess the unique receptor proteins to detect a given combination of MHC and peptide. Secondly, this unique component must be associated with another component common to all T lymphocytes, which is in charge of transmitting the information to the nucleus so that the lymphocyte is activated and reacts to the threat it has detected: a "bad face" in one of our cells indicating an infection, or a cancerous transformation.

The common component of each T-lymphocyte receptor, which is responsible for transmitting information to the nucleus, consists of four different molecules that pass through the cell membrane. Three of them (called with the Greek letters δ, γ and ε) possess each an immunoglobulin domain that is located at the exterior of the cell and forms the denominated **CD3 complex**. The fourth chain is called the zeta chain (represented by the Greek letter ζ) and has a small region outside the cell and a longer region inside that is very important for triggering the internal mechanisms that transmit information to the nucleus. The following figure represents the structure of the T-lymphocyte receptor.

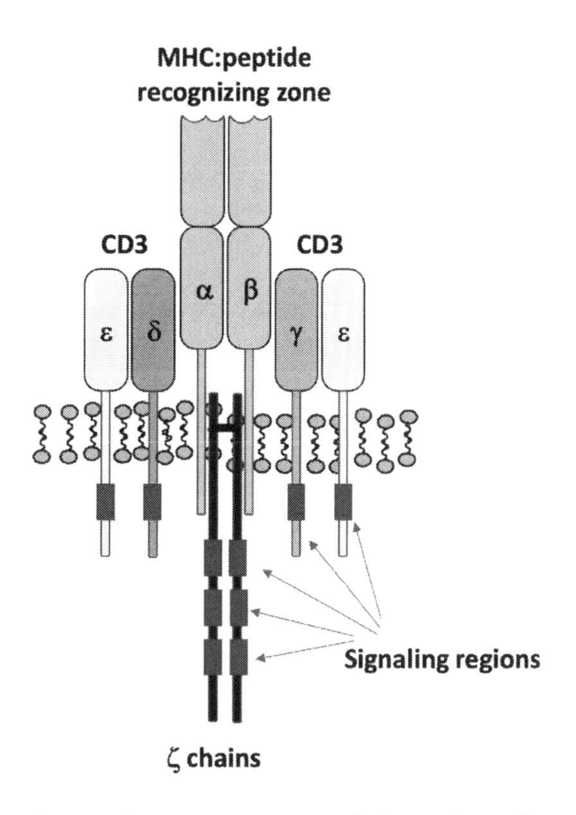


The figure also shows the component of the T-lymphocyte receptor that is different in each lymphocyte. This is formed by the binding of two different protein chains, called α and β. Each of these chains has two immunoglobulin domains, one variable and one constant. We have already seen this combination of immunoglobulin domains in the light and heavy chains of the antibodies, a combination that made it possible for the two variable domains to be able, together, to detect any molecular structure in the outside world. Well, the same molecular philosophy appears here again, with the difference that the T-lymphocyte α and β chains of the T-cell receptor are in charge of detecting changes in molecular structures of our inner world: MHC:peptides. Thus, we now see more clearly a division of functions between the two classes of lymphocytes: B lymphocytes are in charge of detecting external threats, while T lymphocytes are in charge of detecting threats that have somehow changed the cells internally. These can be microorganisms introduced into our cells, such as mycobacteria or viruses, or cells that have mutated internally into tumors.

Having said that, it may not have escaped your attention that the structure of the α and β chains of the T-cell receptor is very similar to the structure of the Fab fragment of the antibodies, which are the ones that possess the binding sites to the epitope of an antigen. In fact, as we have already mentioned, the α and β chains of the T-cell receptors possess a constant immunoglobulin domain and another variable domain, just like the chains that form the Fab fragment of immunoglobulins. As shown in the figure above, the combination of the two variable domains of the α chain and the β chain form the binding site to a peptide presented by an MHC-1 or MHC-2 molecule. Recall that the T lymphocytes that detected each of the MHC molecule types were also of different kind and were called CD8 and CD4 T lymphocytes, respectively.

Thus, the T lymphocyte receptor α and β chains play a role similar to that of the light and heavy chains of antibodies. In fact, the α chain is similar to the light chain and the β chain is similar to the heavy chain. How do we know this, considering in this case both chains are very similar in size?

Scientists have found this out after analyzing the genes that will generate the two chains in each T lymphocyte. As in the case of antibodies, the diversity in the variable domain comes from a process of recombination between gene fragments **V**, **D**, **J** and **C**. Recall that, in the case of antibodies, the light chain genes possessed only fragments **V**, **J** and **C**, while the heavy chain genes possessed, in addition to them, additional fragments of diversity, called **D** fragments. Well, in the case of the genes of the α chain they possess only fragments **V**, **J** and **C**, while the genes of the β chain also possess fragments **D**. This indicates that the α chain is genetically related to the light chain of the antibodies, while the β chain is genetically related to the heavy chain.

The immature, i.e. not rearranged, human α chain gene has 70 to 80 **V** regions, depending on the individuals, i.e. there is a certain diversity in this aspect in the human genome. In addition to the **V** regions, it contains 61 **J** regions, a number that seems to be, in this case, constant among individuals. Finally, the α chain gene has a **C** region that corresponds to the α chain constant region.

The case of the β chain gene is somewhat different, as we do not really have a single gene, but one and a half, with a first half that is shared by two different final halves. The first half corresponds to the region with the **V** fragments, of which there are 52 in this case, apparently without variability in this number between individuals. After this first half of the gene we then have two other halves, one of which, at random, will complete the mature gene. The first half contains a single **D** fragment, 6 **J** fragments and a **C** fragment. The second half contains a single **D** fragment, 7 **J** fragments and another **C** fragment. This is shown in the figure below.

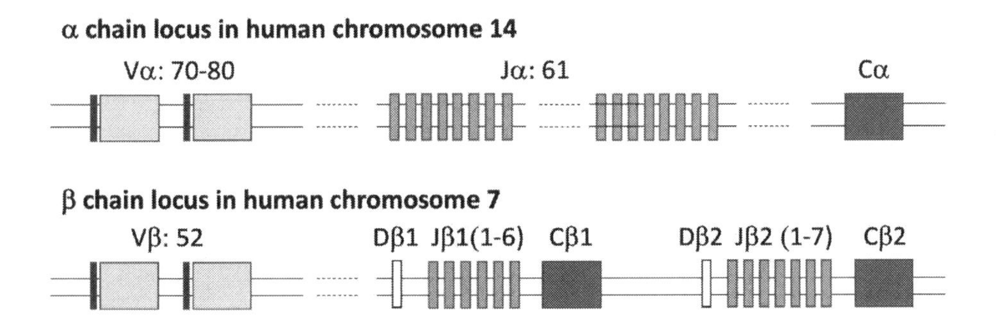


Mature genes will be generated from these immature genes through the process of rearrangement that involves cutting and pasting the DNA as we have seen happen in the case of antibody genes. In fact, the rearrangement is carried out with the participation of the same enzymes that cut and paste the DNA in the B lymphocytes.

In the case of the α chain gene, a **V** fragment is going to be joined to a **J** fragment, at random, which will already generate the mature gene. This will be transcribed, that is to say, it will form a messenger RNA, which will undergo the maturation process that will eliminate the part that separates the united **VJ** region from the **C** region. The mature RNA will then be translated into protein.

The immature β chain gene allows two different mature chain genes to be generated, although each lymphocyte will only generate one at random. In this case, the first recombination occurs between a **D** and **J** region, which join to form a **DJ** fragment. This union can happen between the **$D_{\beta1}$** region and any **$J_{\beta1}$** region or between the **$D_{\beta2}$** region and any **$J_{\beta2}$** region. Any **V** region will join this region to form the mature gene, which will be either **$VD_{\beta1}J_{\beta1}$** or **$VD_{\beta2}J_{\beta2}$**. Subsequent transcription from these genes generates an immature messenger RNA that is processed to attach the corresponding **$C_{\beta1}$** o **$C_{\beta2}$** region to the **VDJ** fragment and generate a messenger RNA that will be translated into a **β1** or **β2** chain. The figure below represents an example of recombination between the fragments that make up the α locus and the fragments that make up the β locus to generate mature genes that are translated into the protein chains of the TCR receptor.

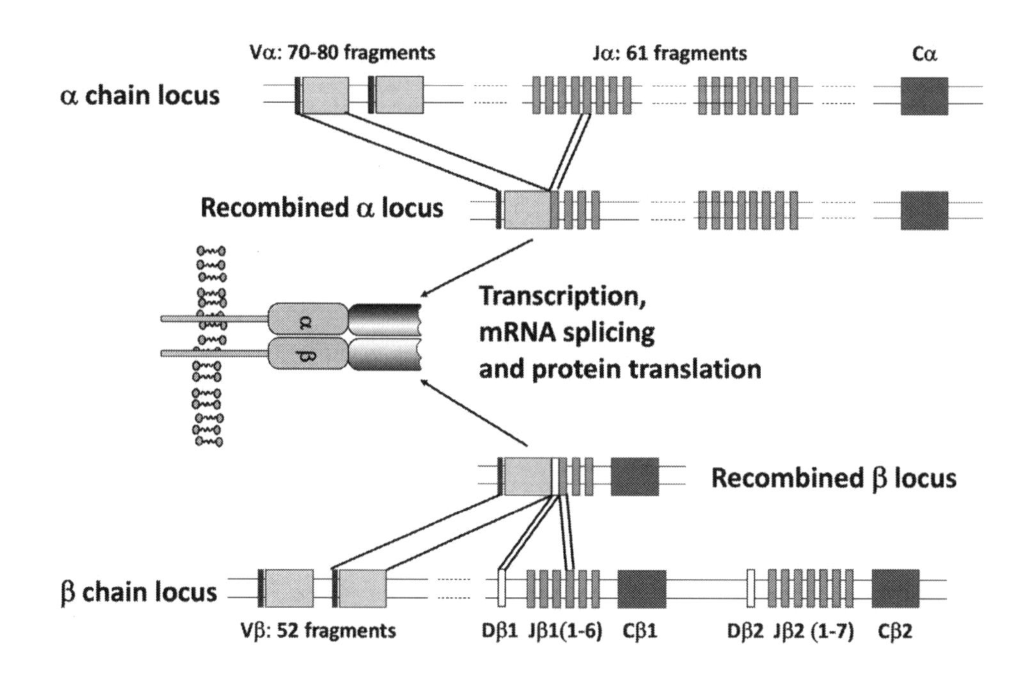


In this way, the gene recombinations that occur in each T lymphocyte generate an α chain of one type only, and a β chain that can be either type β1 or type β2. The combination of the α chain with whatever β chain generated produces the region of the receptor capable of interacting with a MHC molecule and its peptide.

## 5.4.- SIMILARITIES AND DIVERGENCES BETWEEN T AND B RECEPTORS

There are many similarities in the process of generating T-cell and B-cell receptors; the latter are also converted into the secreted antibodies. However, there are also significant differences between the two processes. The first difference is that T-cell receptors do not have an Fc region. The logical reason for this difference is that T receptors do not have to be secreted externally to perform their functions, as these are confined to the T cell. However, the antibodies produced by B lymphocytes, derived from the B-cell receptor, are destined for the external world. Once secreted by the activated B cell, it no longer has to perform any additional functions. On the contrary, these additional functions now depend on other molecules or cells that detect the Fc regions of the various antigen-bound antibody classes.

The second fundamental difference between T-lymphocyte receptors and B-lymphocyte receptors is that the former are restricted, while the latter are not. What does this restriction mean? Well, it means that T-lymphocyte receptors have their limitations, while B-cell receptors and antibodies have not. What limitations are we talking about? Of course, we're talking about limitations on what these molecules can or cannot do. The receptors for both classes of lymphocytes are intended to detect molecules outside the cells that produce them. However, B-cell receptors, as we have already said, can detect any molecular structure present on the surface of molecules, but T-cell receptors are restricted to detect MHC-1 or MHC-2 molecules with bound peptides.

The restriction of detecting only MHC molecules of one type or another, never both at the same time, is a truly extraordinary fact. Although mature T receptor genes are formed by the random joining of DNA regions, as we have already seen, the **V**, **D** and **J** regions, and although these regions are diverse, their diversity is not infinite and the amino acid sequences of the proteins they generate are designed to interact with the amino acids and the shape of MHC molecules.

The most variable part of the MHC:peptide complex with which T-cell receptor molecules interact is, of course, the peptide. The MHC molecules themselves are less variable, because they come from our own genes, and each person has only a few of them. For this reason, genes that produce T-cell receptors may have been selected throughout evolution to generate proteins that bind to MHC proteins. Restriction would therefore refer to this setting. However, the receptors should not be restricted to thus enable them to detect the billions of different peptides that can be presented by MHC molecules. In this case, the restriction is not appropriate, among other things because the peptides are very diverse and come from both our own proteins and foreign proteins. Restricting the receptors so that they only detect certain peptides and not others could lead to some foreign peptides escaping detection.

The solution to this contradiction is provided by the gene recombination mechanism itself. We should remember that, in the case of antibodies, the greatest diversity was produced in the CDR3 region **(section 4.3)**, since this came from the union of the **VJ** or **DJ** fragments, depending on whether the chain was the light or the heavy one. As in the case of antibodies, T-lymphocyte receptors also have three regions of complementarity per chain, **CDR1**, **CDR2** and **CDR3**. The information for the amino acid sequence of the **CDR1** and **CDR2** regions is already present in the **V** regions of the genome, i.e. it does not change in the recombination process. However, the amino acid sequence of the CDR3 region also comes from the **VJ** or **DJ** recombination, depending on whether the chain is the $\alpha$ or the $\beta$ chain. This means that the CDR3 region of the T-cell receptor chains is much more diverse than the CDR1 and CDR2 regions.

What is extraordinary about this is that studies to find out which regions of the T-lymphocyte receptor interact with MHC molecules with their bound peptide have revealed that the CDR1 and CDR2 regions contact the MHC-1 $\alpha$ chains or the MHC-2 $\alpha$ or $\beta$ chains. However, the amino acids of the CDR3 region contact with the amino acids of the peptides that the MHC molecules carry attached. All three interactions are necessary for a T-cell receptor to detect and bind to an MHC molecule. The links formed between the CDR3 region of the T-cell receptor and the peptide are essential to achieve a stable interaction, but it is equally true that the links between the CDR1 and CDR2 regions with the MHC protein chains are also necessary. If either type of interaction fails, the T-cell receptor cannot bind strongly to the MHC molecule. Thus we see that all T-cell receptors are "designed" to interact with MHC molecules, thanks to the amino acids that are derived from the **V** sequences present in the genome, but this interaction is only possible if these molecules carry a peptide that binds to the CDR3 region of a particular receptor or, otherwise, a stable interaction will not occur. Fortunately, the latter region is not restricted in its amino acid sequence, since it is generated in the recombination process. These limitations do not apply to B-lymphocyte receptors, which, although they also possess amino acids derived from **V** regions present in the genome, do not have these regions restricted in such a way that the amino acids derived from their sequences preferentially interact with any other molecule, but rather have complete freedom to detect any molecule.

## 5.5.- It's going to be CD4 or CD8?

The restriction that limits T lymphocytes to interact only with MHC molecules is deeper than it appears. We have seen above that T lymphocytes are of two main classes: CD4 T lymphocytes and CD8 T lymphocytes. Well, CD4 T lymphocytes are restricted to interact only with MHC-2 molecules and their bound peptides, while CD8 T lymphocytes are restricted to interact with MHC-1 molecules and their peptides. During maturation of T lymphocytes, they must be converted to either CD4 or CD8 T cells. What signals tell a developing lymphocyte, which has successfully rearranged its T-cell receptor genes, whether it should become a CD4 or a CD8 T lymphocyte? Where do the lymphocytes receive these signals?

The conversion of T lymphocyte precursors into mature lymphocytes is such an important process for the survival of the organism that it has a dedicated organ. This organ is the thymus, which we briefly discussed at the beginning. The thymus is organized into two identical lobes, each of which is divided into two zones called the **medulla** and the **cortex**, which play different but complementary roles in the development of T lymphocytes. The marrow and cortex cells provide the T lymphocyte precursors with the various signals needed to convert them to mature T lymphocytes of a particular class.

The thymus has its own cells, called **stromal cells**, which produce chemokines that are attractive to T lymphocyte precursor cells as well as other cells of immune origin. The cells that are attracted to the stromal thymus cells are called **thymus hematopoietic cells** because they are derived from the bone marrow from which all blood cells are generated in an overall process called **hematopoiesis**.

Thus, in the bone marrow, the lymphoid stem cells are becoming immature daughter cells. In this conversion, some of the immature daughter cells will mature into B lymphocytes. In this case, the entire maturation process occurs in the bone marrow, from where the generated B lymphocytes exit into the blood and travel to the lymph nodes in search of potential antigens. However, many of the daughter cells become precursors to T lymphocytes and leave the bone marrow to the bloodstream before maturing. These cells are those attracted by the chemokines produced by the thymus stromal cells. When these cells pass through the blood capillaries of the thymus and detect chemokines, they attach to the endothelium of these capillaries, pass through them, and settle in the thymus. Once in the thymus, these cells will continue their maturation and will undergo the processes of positive and negative selection which will ensure that the mature T lymphocytes generated will detect MHC-1 or MHC-2 molecules with sufficient intensity not to activate themselves intensely against them, in the hope that perhaps in the future a foreign peptide attached to these molecules would make it necessary to activate them and participate in the defense mechanisms.

The precursors of the T lymphocytes that reach the thymus are called **thymocytes**. These cells have not yet rearranged the genes for the T cell receptors. This rearrangement begins when the thymocytes receive signals from the thymus stromal cells that tell them they are in the right place and can begin to mature.

In addition to sending signals for thymocytes to begin rearranging their $\alpha$ and $\beta$ chain T cell receptor genes, thymus stromal cells also display MHC-1 and MHC-2 molecules on their surface that carry self-peptides and can stimulate the receptors of thymocytes as they generate them. Thymus stromal cells are very interesting from the point of view that they can produce practically all the proteins in the body, generating peptides from them and presenting them in MHC-1 molecules. Let us bear in mind that each organ produces only the proteins that are necessary for its function. For example, many proteins produced by the liver are not produced by the heart, or by the intestine. However, the thymus must perform the function of showing the T cells that are making "masks" all the possible "faces" of the body in which those T cells are being generated, and this is what it does. At the same time, the stromal cells of the thymus also take up their own proteins from blood and extracellular fluids, degrade them and display the peptides derived from them in the MHC-2. The expression of MHC-2 in these cells is also exceptional, since only antigen-presenting cells, not other cells in the body, express MHC-2. Thymus stromal cells, therefore, express and present peptides of almost all the self-proteins, both in MHC-1 and MHC-2 molecules.

Thus, as soon as the thymocytes have successfully rearranged their T-cell receptor genes, they can check whether the receptor produced is functional and suitable for detecting our own "faces". The detection of self "faces", let us remember, is necessary to ensure that the receptor is going to be useful, i.e., it is a well-formed "mask". When a thymocyte detects with its newly formed receptor an MHC molecule with a self-peptide, so that the interaction produced is not very strong, it receives a survival signal that prevents the cell from committing suicide by apoptosis. Thymocytes that have not successfully rearranged their receptor genes and cannot detect even faintly an MHC molecule and its peptide do not receive these survival signals and die by apoptosis.

Obviously, after the random rearrangement of the receptor genes, receptors can be produced that bind preferentially to MHC-1 molecules or that bind preferentially to MHC-2 molecules. In the first case, thymocytes are converted to CD8 T lymphocytes, and in the second case they are converted to CD4 T lymphocytes. This conversion involves other signals besides those received by the T-cell receptor, but the most important signal is that sent by the receptor that detects an MHC-1 or MHC-2 molecule.

Let us remember at this point that the interaction of the T-cell receptor with an MHC molecule of one type or another requires the CD4 or CD8 co-receptor molecules, which collaborate with the receptor to bind to the MHC molecules forming a molecular "clamp". Without the presence of these molecules, lymphocytes cannot interact with the MHC molecules, as they are missing one of the parts of the gripper. For this reason, at one point in their development, when they have already rearranged the T-cell receptor genes, thymocytes produce and express on their surface the two molecules, CD4 and CD8. In this way, whatever kind of MHC molecule its receptor prefers to interact with, the necessary "clamp" can be formed.

However, the result of forming a "clamp" with a CD8 molecule is different from that of forming a "clamp" with a CD4 molecule. CD4 and CD8 act as co-receptors, i.e. they work together with the T-cell receptor to send a signal into the cell and switch genes on or off, which is almost always what molecular signals do. One of the consequences of this signal is that, if the "clamp" is formed with a CD8 molecule, i.e. if the receptor binds to an MHC-1 molecule, then the signal sent through the receptor and through CD8 ends up silencing the activity of the CD4 gene, so that the CD4 molecule stops expressing itself on the membrane. The cell thus becomes a mature CD8 T cell that can only be activated by foreign peptides presented by MHC-1 molecules. On the other hand, if the pincer is formed with the CD4 co-receptor, the signal sent through the receptor and in this case of CD4 ends up silencing the gene that produces CD8, with which the cell will become a mature CD4 T cell that can only be activated by foreign peptides presented by MHC-2 molecules.

Thymocytes that develop into mature CD4 or CD8 T lymphocytes have therefore passed a selection process, called **positive selection**. This process, as we have said, ensures that the lymphocytes formed interact with our own identity "faces". However, it is possible that this interaction is sometimes too strong. We have explained before that, if that happens, the generated T lymphocyte could attack the organism itself, which would generate autoimmune diseases. To prevent this, lymphocytes that interact too strongly with MHC molecules receive an intense signal that also induces cell death by apoptosis. This process is called **negative selection**, because it eliminates (denies) the existence of lymphocytes that can react too strongly against our own MHC molecules loaded with our own peptides. The combination of the processes of positive and negative selection ends up generating a repertoire of mature naïve T lymphocytes guaranteed of having well-formed T-cell receptors, capable of detecting self MHC-1 or MHC-2 molecules charged with their corresponding foreign peptides and, at the same time, unable of strongly reacting against self MHC molecules loaded with self-peptides. These T lymphocytes will only be activated when self MHC molecules present them with some foreign peptides that have not been presented in the thymus during the process of generating this repertoire of mature naive T lymphocytes.

At this point, we can end our journey through the basics of the immune system. We will now look at other interesting aspects of this system that affect the lives of millions of people every day, in particular vaccines, which can save us from premature death, and we will also explore the very interesting mechanisms that some microorganisms use to evade the attack by the immune system, which may explain why many infectious diseases lack an effective vaccine. Finally, we will address the issue of how the immune system defends itself against new emerging viruses, such as the coronavirus SARS-CoV-2, and try to explain, based on what we already know, why some people react asymptomatically to this virus and others succumb to its infection. This difference is not related to the virus itself but to how people's immune system reacts to it, although there may be other factors as well. In addition, when dealing with these issues, we will take the opportunity to introduce new concepts and actors of the immune system when necessary, which will allow us to go deeper into its wonderful functioning. Go ahead then with courage, because what follows is only accessible to the brave intellectuals, the very curious, and those in love with science.

## 6.- VACCINES AND OUR HEALTH

With what we have learned up to this point, we can better appreciate the contribution that the invention of vaccines has made to humanity. Despite what one might think in developed countries, where cancer and cardiovascular disease are the leading causes of death, infectious diseases are the leading cause of death worldwide. The two most important contributions to public health so far have been the improvement of health conditions, including sanitation, and vaccination. Both together have reduced deaths from infectious diseases very significantly. Thanks to vaccination, smallpox was eradicated in 1980, one of the most outstanding achievements of modern medicine. Today, vaccination is bringing us closer to eradicating another important disease: polio.

Vaccination aims at inducing an adequate protective immune reaction against one given pathogenic microorganism. The word "adequate" is important. We have already seen that the immune system identifies the type of threat it must deal with and makes the decisions it considers most effective in combating it. Therefore, vaccines must be able to induce the appropriate responses in the immune system to protect us from the microorganism against which we vaccinate.

For example, if the pathogenic microorganism is to be combated by opsonization and phagocytosis, the vaccine needs to induce the production of opsonizing antibodies, such as antibodies of the IgG1 class, but not antibodies of other classes. If the vaccine induces antibodies, but not of the correct class, as would be the case if it induced antibodies of the IgE class, it would not be effective in protecting us and could also cause allergies. This implies that the components of a vaccine must activate dendritic cells or antigen-presenting macrophages in the same way that they would be activated by the genuine pathogenic microorganism against which we wish to be vaccinated. In this way, these cells will activate the T lymphocytes and later the B lymphocytes in the correct way.

However, the benefits of vaccination not only depend on it inducing the right immune mechanisms, but also on it inducing the so-called

**immune memory**. Immune memory is a fundamental phenomenon for the protection that the immune system must provide to organisms, and without it we would be much more susceptible to disease, not to mention that vaccines would be impossible.

Once the adaptive immune system has overcome an infection, the effector cells remove the antigens and molecules of the pathogen that initially activated them and attracted them to the site of infection. In the absence of the continuous stimulus provided by the presence of antigens, after the elimination of the infectious organisms, the effector cells no longer receive survival signals, i.e. signals that inform them that they are still needed, and die of apoptosis, after which their remains are rapidly eliminated by phagocytes. These cells, in addition to phagocyting microorganisms, also have the function of ingesting and digesting the dead cells. They can do this because phagocytes have receptors on their surface that recognize the lipid **phosphatidylserine** on the outer membrane of apoptotic cells. This lipid is actively maintained in the inner layer of the cell membrane, that is, it is necessary that a cell be alive and healthy for it to be able to actively maintain this lipid, using energy, in the inner layer of the cytoplasmic membrane. Only when the cell starts dying, this lipid appears in the outer layer of the membrane and constitutes a molecular signal of the type "**eat me**". These signals are identified by the phagocytes to eliminate the dead cells by phagocytosis. In this way, the elimination of the infection leads not only to the elimination of the pathogen, but also to the disappearance of the effector cells specific to the latter.

However, not all activated lymphocytes become effector cells and die once the infection is cleared. Some activated lymphocytes become **memory cells** in the process of activation. Some of these cells remain alive throughout the life of the organisms and provide long-lasting immunity against diseases as serious as measles. How are these memory cells generated and what characteristics do they possess?

Not surprisingly, there are memory cells of the two main types of those that constitute adaptive immunity, namely memory B cells and memory T cells. Research into their generation process and characteristics has not been easy, as these cells are produced in much smaller numbers than effector cells. However, the most recent studies have managed to reveal both their generation process and their main characteristics and the differences that these cells show with naïve cells and with effector cells of the same class.

Later on we will talk about these differences which, as it shouldn't be surprising, result from the differential functioning of different genes that make possible their capacities as memory cells, among which we can find the capacity to be activated much faster than the naive cells after the encounter with the same antigen that led to their generation. Let's talk first, however, about the process of their production, which is not yet fully known in all its details. We will begin with the process of producing B memory cells.

## 6.1.- Generation of memory B cells

As we know, effective protection against many microorganisms, particularly viruses, depends on the secretion of neutralizing antibodies against them. Humoral immune memory therefore depends on the presence of these antibodies in sufficient quantities, even in the absence of infection, as well as on the generation of high quantities of specific antibodies in the event that infection cannot be prevented by pre-existing antibodies.

Antibodies, like all the proteins found in the blood and the internal body fluids that bathe the cells, age gradually, lose their antigen-binding properties over time and are eliminated at a certain rate. This implies that, to maintain a constant concentration of these molecules, they must also be produced constantly. This constant production of protective antibodies is carried out by long-lived B cells, located in the blood plasma, which continuously produce antibodies against the antigen first encountered by the naïve B cell from which they were derived. These continuous antibody-producing plasma B cells must be generated, in the course of the battle against an infection, in appropriate quantities to provide adequate protection against a second encounter with the infectious microorganism.

However, the generation of these cells is not enough to ensure adequate protection. Long-lived B cells that do not produce antibodies must also be generated, but if they find their antigen again, they must be able to activate rapidly and become highly antibody-producing cells. As mentioned above, the antibodies produced by one class of memory cells or the other must also be of the right class to neutralize or opsonize the microorganism in question, by activating complement or inducing direct phagocytosis by the phagocytes. This implies that at least some memory B cells should be derived from those that have made the change of antibody class and thus be prepared to immediately produce the antibody of the correct class. Others, however, may undergo class switch in subsequent encounters with the same antigen. Similarly, it would be desirable that the antibody generated should be of high affinity, i.e. some memory B cells should be derived from an activated B cell that has undergone the process of somatic hypermutation. Others, however, may also undergo somatic hypermutation after activation in subsequent encounters with their antigen.

Studies have confirmed that a second encounter with the same previous microorganism leads to the generation of antibodies mostly of the class that was generated against it after the first encounter, although antibodies of the IgM class can also be generated, which, let us remember, is the first class of antibodies secreted by the activated B cells. This indicates that there are memory B cells that have undergone class switch and others that have not yet done so. The average affinity of the antibodies generated is, however, also generally higher than that generated at the first encounter with the antigen, and this affinity may even increase after a third or subsequent encounters with the same microorganism. This seems to indicate that some memory cells have undergone the process of somatic hypermutation before becoming memory cells and that others may undergo it in a subsequent encounter with their antigen.

To make this possible, therefore, after the activation of a B cell by an antigen and the generation of a clone of activated B cells, some of these cells must be converted into memory B cells of various types, which together provide much safer protection against the microorganism that caused an infectious disease after a first encounter with it, a disease that was fortunately overcome thanks to the initial, or primary, immune response. To understand the process of generation of these cells, we must visit again, with a little more depth, the generation of activated B cells after their encounter with an antigen. Remember that for the correct activation of B cells and for the immunoglobulin class switch it is necessary that at least a T cell clone collaborates with the B cells by sending it both secreted cytokines and membrane stimulating ligands. This collaboration takes place in one lymph node, of which we have several hundred distributed throughout the body.

Lymph nodes are organized structures whose function is to trigger the adaptive immune response efficiently. The lymph node structure consists of an outer layer, which is like the wall of a vessel, and an inner marrow, which is the contents of that vessel. Arteries and veins enter and leave that organ, respectively, as do afferent (ingoing) lymphatic vessels and efferent (outgoing) lymphatic vessels. The afferent lymphatic vessels supply lymph and antigen-presenting cells from the periphery of the body, that is, from areas where sites of infection may develop. The blood supplies naïve B and T lymphocytes, which enter the lymph node by the process of extravasation, and which, in the absence of an encounter with the antigen, will leave the node via the efferent lymph vessels, from where they will return to the blood via the thoracic duct.

Once inside the ganglion, the lymphocytes are organized into specific areas, which are called **T zones**, formed not surprisingly by T cells, and **B zones**, formed also not surprisingly by B cells. The organization into zones occupied by each class of lymphocyte is possible because T cells are attracted to their zone by certain chemokines, just as B cells are attracted to their zone by other chemokines to which these cells respond. These chemokines are secreted by cells of the stroma, that is, the structural skeleton of the lymph node.

The B cells inside the ganglion form the so-called **lymphoid follicles**, which are located outside the medullary region. These lymphoid follicles are in contact with the T-cell areas, in the **paracortical areas** located towards the interior of the organ. B and T cells, therefore, have an interaction surface that is equivalent to the border between both T and B zones. This interaction boundary is fundamental for the correct activation of B cells, because it is at this boundary that they will find a helper T cell that has been activated by the same antigen as them.

T cell areas are also populated by dendritic cells that reach the lymph node from the periphery of the body, as we have said. These cells present antigens to the T cells. If any T cell has a receptor that binds strongly to the foreign peptides presented, for example, by MHC-2 exposed on the surface of dendritic cells (in this case it is a matter of activating a T cell that will help B cells), the T cell can be activated to an effector $T_{FH}$ cell, capable of providing stimulating signals to B cells.

On the other hand, the lymph node B cells can also be activated by antigens that arrive there. These antigens are transported to the node by the lymph, and, in addition, are retained there by dendritic cells different from classical dendritic cells, called **follicular dendritic cells**. These cells are not derived from the bone marrow, as classical dendritic cells are, but are derived from a tissue called mesenchyme. Follicular dendritic cells do not migrate through the lymph. They reside in the lymph node where their function is to store in their membrane the antigens that arrive at the node transported by the lymph to show them to the B cells. Their function is fundamental in the process of somatic hypermutation, since it is these dendritic cells that allow B cells to check whether the mutations that they have generated in the immunoglobulin genes have led to the generation of receptors with a greater affinity for their antigen. If this were the case, remember **(section 2.5.3)**, then the cells that are descended from the original one will be able to receive signals through their improved version of their receptor that will allow them to remain alive. If, on the other hand, the mutations have not generated receptors of greater affinity, they will not receive these signals and the cells will die by apoptosis, as they will not be able to compete successfully with the other B cells for binding to the antigen that the follicular dendritic cells present to them.

Let us also remember that before B cells undergo the process of hypermutation, those that detect an antigen given by one of their epitopes incorporate it into their interior by endocytosis and digest it into peptides that are presented by MHC-2 molecules. In addition, the detection of the antigen sends a first activation signal to the interior of the B cells allowing them to acquire sensitivity to certain chemokines that direct them to the border between the T zone and the B zone of the lymphoid follicle.

$T_{FH}$ cells that have been activated by dendritic cells in the T-cell zone are also sensitive to these chemokines and are equally directed by them to the boundary between T and B zones. We now see that the organization of the lymph nodes into zones of cells separated according to their class is very important to facilitate the interaction between the activated T- and B-zone cells. These are first activated individually and specifically by different mechanisms in their respective zones of residence, after which they are directed to a specific meeting point, which is not really a point, but a meeting surface, through which the different activated cells glide and touch each other in search of the companion cell that has been activated by the same antigen. In this way, from a set of cells that are in a three-dimensional space in which it would be very difficult for two specific cells activated by the same antigen to meet, the probability of this encounter is greatly increased, firstly because only activated cells will come to this contact surface and, secondly, because their distribution is reduced to a two-dimensional space, in which it is much easier for T and B cells activated by the same antigen to meet.

Once this encounter has taken place, the $T_{FH}$ cell will supply signals to the B cell that will lead to its rapid proliferation and the generation of a clone of identical B cells. At the same time, the $T_{FH}$ cell itself will also reproduce, thanks, in part, to the signals sent through the MHC-2:antigenic-peptide complexes presented by the B cells and also thanks to the co-stimulatory molecules expressed by the B cells on their membrane. This happens in only two days after the first encounter between T and B cells.

On the third day, some of the B cells are transformed into cells called **plasmablasts**, which travel from the B-cell area to another region of the lymph node called the **subcapsular sinus**. In this region, these cells already begin to secrete high amounts of IgM class antibodies, as the class switch has not yet occurred. Plasmablasts are short-lived cells that do not allow sustained production of antibodies, which, as they are mainly of the IgM class, will only activate complement, as they are unable to effectively neutralize antigens or opsonize them on their own. However, some of these plasmablasts can also perform the class switching process and generate other antibody classes, instead of IgM.

On the other hand, a few of these B cells are going to be transformed into B memory cells. These memory B cells will enjoy a long life and will be able to detect again the same original epitope that led to their formation. Then they will become activated again to repeat the process we have just described and produce antibodies. These memory cells, however, as they come from the original activated B cells that have not undergone class switch or the somatic hypermutation processes, will not, in principle, produce antibodies optimally effective against a second reinfection.

Fortunately, most of the activated B cells on the border between the B and T zones will migrate back to their B cell zone and reach the interior, where the follicular dendritic cells are located. The activated $T_{FH}$ cells that found a companion B cell activated by the same antigen at that border will also migrate, but not to their T cell zone. They will instead accompany the activated B cells into the B cell zone. The three cell types will interact and form inside the lymphoid follicle the so-called **germinal center**.

In the germinal center two very important processes will take place: antibody class switch and somatic hypermutation. The latter process is organized thanks to the fact that the germinal center is structured into two different zones, the so-called **light zone** and **dark zone**. In the dark zone, the B cells are very packed, very close to each other, which is why under the microscope this region of the germinal center looks dark. There B cells are reproducing and generating immunoglobulin receptor gene mutant daughter cells. Once these mutants have been generated in a first round of cell division, the B cells leave the dark zone and, following a gradient of concentration of specific chemokines, they head towards the light zone. It is in this zone where the follicular dendritic cells that have accumulated antigens transported by the lymph from the site of infection, or from the vaccination site, reside. These antigens are presented to the mutant B cells, which thus check whether they possess a receptor with sufficient affinity for the antigen. If they do not, they die by apoptosis, but if they do, they receive signals that induce them to reproduce again, to travel again back to the dark zone of the germinal center and to mutate their B cell receptor genes once more. After this second round of mutation, the descendants of the first cells travel back to the light zone to check again the affinity of their receptors. Again, only the cells with receptor of the highest affinity survive, being able to successfully compete for the always limited amount of antigen with their less fortunate mutated B-cell sisters.

After several rounds of reproduction and mutation, the B cells leave the germinal center and head for the bone marrow, where they will become mature antibody-producing B cells. However, in each round of mutation and selection, some of the surviving cells become memory B cells. These will be the ones that in a second encounter with the antigen will react more quickly to generate antibodies against it.

We thus see that this process generates not one, not two, but a whole population of memory B cells, no longer against one antigen, but against each of the epitopes of that antigen, against which they will react with a range of affinities. It is important to note that to generate these memory B cell populations only the collaboration of the same $T_{FH}$ cells is necessary, that is, it is not necessary to generate different $T_{FH}$ cell clones that react against different peptides presented on MHC-2 molecules by the B cells. A clone of $T_{FH}$ cells that react against a single MHC-2:peptide is enough. However, it is possible that more than one T cell can be stimulated by a classical dendritic cell and thus generate several different clones of $T_{FH}$ cells, all of them capable of collaborating with the B cells that have reacted against the same antigen.

The different B memory cells, which make up a given population, are cells that will react with different affinities to the same epitopes. This allows an interesting phenomenon that is that, if the antigen mutates and generates slightly different epitopes, some of the memory B cells will still be able to react to it, even with greater affinity than the initial one, in some cases. In other words, B cells that initially react with low affinity against an epitope can generate by somatic hypermutation cells with higher affinity receptors against it. Some mutant pathogens in that epitope may prevent B cells and antibodies from neutralizing them. However, it is likely that, having mutated that epitope, it is now the low affinity antibodies against it that acquire high affinity, not because the antibody has mutated, but because the epitope has mutated. In this way, the generation of a population of memory B cells of different affinities allows a much better protection against future mutant potentials of the same microorganism that will need to be neutralized.

## 6.2.- Generation of memory T cells

The generation of memory T cells is even more complex than that of memory B cells. To begin with, there are memory CD8 T cells and memory CD4 T cells. In addition, of the latter, as we know, there are several classes, such as $T_H1$, $T_H17$, etc. Each of these classes must reside in a particular tissue, such as the intestine, or in a lymph node. This implies that each type of memory T cell must possess the necessary molecules to enable it to address and reside in the appropriate tissue. It is in this tissue that the memory T cells can perform their function most effectively, as it is usually in the tissue in which they reside where they must be activated again when they find the antigen that first activated the naïve cells from which they were derived. For example, memory $T_{FH}$ cells must reside in or around the lymphoid follicles to provide the right signals to the B cells so that they can effectively produce antibodies.

### 6.2.1.- Memory CD8 T cells

Let's first explain the generation of memory CD8 T cells, which is the simplest. In previous sections we have explained that naïve CD8 T cells are activated to effector cytotoxic CD8 T cells, which are mainly involved in the elimination of virus-infected cells. Well, elegant experiments have shown that the first thing that happens when a dendritic cell presents an antigen to a naive CD8 T cell is that it first becomes one of two types of memory cell. The first is called a **stem memory CD8 T cell ($T_{MM}$)** and the second is known as a **central memory CD8 T cell ($T_{MC}$)**. Both types of cells divide slowly and are long-lived. If these cells continue to receive signals, from the antigen presented by the dendritic cells, some of them become activated and divide to give rise to a larger population of **memory precursor CD8 T cells ($T_{MP}$)**, which are in effect the precursors of the **effector memory CD8 T cells ($T_{EF}$)**. The memory precursor cells can be converted, if appropriately stimulated by antigen and co-stimulator molecules, into the actual cytotoxic effector CD8 T cells, which are the main cells responsible for inducing apoptosis in virus-infected cells. These CD8 effector T cells will also die of apoptosis once their mission is completed.

Stem memory CD8 T cells and central memory CD8 T cells reside in the lymphoid organs, although they can also be found circulating in the blood and lymph, probably recirculating through the different lymph nodes of the body as do naive T cells in search of their antigens. However, precursor memory CD8 T cells derived from central memory cells can be found, in addition to lymphoid tissues and blood, in non-lymphoid organs and tissues, where they can be activated by dendritic cells resident in those tissues. Thus, these cells are activated *in situ* as soon as the antigen has been detected a second time. This saves time, because the dendritic cells do not have to travel through the lymph to the lymphoid organs and find the T cells there to activate them. However, it is not safe to do this with naive cells, but only with memory cells, which the immune system already knows come from a previous encounter with a microorganism that has been eradicated thanks to the activation of the same type of cells that are now activated again in response to a second encounter with the same microorganism. Overly "cheerful" activation of naïve CD8 T cells could eventually cause serious autoimmunity problems and be more harmful than beneficial for that reason.

Not surprisingly, activation of precursor memory CD8 T cells in an environment different from that of a lymphoid organ, where activation of naive T cells occurs, is possible because these precursor memory T cells possess the molecular tools that allow them to be activated more easily under less favorable conditions than those found in the lymph node. Furthermore, these effector memory CD8 T cells resident in tissues are not attracted by chemokines to the lymph nodes nor do they possess the molecules necessary for extravasation in the capillaries of those organs, although they can be extravasated in the blood vessels of non-lymphoid tissues. Later we will briefly discuss the molecular characteristics that differentiate memory cells from naïve cells.

#### 6.2.2.- Memory CD4 T cells

Let's now turn briefly to memory CD4 T cells. Like CD8 T cells, after activation naïve CD4 T cells also generate **memory stem cells** and

**central memory cells**. The latter are not predetermined memory cells for the different subclasses of CD4 T cells. Instead, they recirculate through the lymph nodes as do naïve CD4 T cells. They will be activated and differentiated into effector CD4 T cells of the appropriate class ($T_H1$, $T_{FH}$, $T_H17$, etc.,) according to the cytokines sent by the dendritic cells that present the antigen to them.

There are also **effector memory CD4 T cells** that reside in tissues, although they can also be attracted to sites of infection when one occurs. It is unclear whether these effector memory CD4 T cells are activated and become the appropriate type of CD4 T cell based on signals received from dendritic cells or whether they are already committed to become the type of CD4 T cells that the naïve CD4 T cells from which they derive were induced to differentiate to. In the latter case, these memory CD4 T cells would only require an encounter with antigen and co-stimulator molecules for activation, but they would not necessarily need the cytokines that naïve CD4 T cells do require, since they would already be pre-programmed to differentiate to a particular type of effector T cell.

In any case, we see that both CD4 and CD8 memory T cells are organized in a similar way, with the initial generation of a class of cells that act as specific stem cells of the type of effector T cells that are needed to fight a particular pathogen. This set of stem cells is important for always maintaining a population of T cells in the lymph nodes that are capable of rapidly dividing after a second or subsequent encounter with the antigen and capable of generating effector T cells of the correct class.

## 6.3.- Molecular features of memory cells

Immunological memory is not a mysterious phenomenon. It is based on the generation of a population of long-lived cells, more numerous than that of naïve cells specific for an antigen, which are capable of being activated rapidly if they encounter the same antigen. The higher number of antigen-specific memory cells increases the probability that they will find the antigen very quickly if it tries to invade the body again. As is generally the case with differences between cells, the cause of differences between memory cells and naïve cells lies in the different expression of certain genes. It is these genes that provide the different properties to the different cells.

Let's start by explaining why memory cells are long-lived, whether they're B's or T's. The longevity of all classes of lymphocytes depends on their susceptibility to apoptosis. Death of these cells by apoptosis is a default mechanism that they have from their generation. We have already seen that during the development of B and T lymphocytes, unless they receive survival signals from their antigen receptors, they will die of apoptosis. Those survival signals are the ones that indicate them that they have generated functional receptors and therefore are potentially useful lymphocytes. Therefore, lymphocytes, in general, even those not activated, must continuously receive signals from the environment to stay alive. In the absence of these molecular survival signals, apoptosis is triggered. These survival signals can be generated by a basal activity of the antigen receptor, i.e. by the activity that this receptor possesses even when it has not been activated by a foreign antigen. The cytokines found in the lymphoid organs, where some lymphocytes reside and through which they all recirculate in and out of them while patrolling the body for pathogens, are also important.

After activation of the lymphocytes, they increase the requirement level of survival signals received through the antigen receptor. These signals are received as long as there is antigen present, i.e. as long as the infection has not been completely overcome. When the infection is overcome, the antigen disappears from the body, and the effector cells can no longer receive survival signals at the required level, so apoptosis is triggered.

Memory cells are produced, as we have seen, during the process of lymphocyte activation, but they are different from the effector cells generated in several respects, in particular that they do not depend on the survival signals received through their antigen receptor to survive. The reason for this independence is the increased expression of genes that produce antiapoptotic proteins, which block the process of programmed cell death. There are several such proteins, although perhaps the most important is the **Bcl-2** protein. Memory cells express much higher levels of Bcl-2 than effector cells and naïve cells, although the latter also express higher levels than effector cells, indicating that the latter, when activated, increase their sensitivity to apoptosis, as we have explained.

In addition to having a long lifespan, memory cells have a greater capacity to become active again and generate effector cells. This activation can be done with smaller initial amounts of antigen than those needed to activate naïve cells. In the case of B cells, this lower need of antigen for activation is possible, in part, because the antigen receptors of memory B cells have a greater affinity for it than the receptors of naive cells, which, unlike memory B cells, have not undergone somatic hypermutation. This allows the memory B cell antigen receptors to pick up antigens more strongly, which makes it easier for the receptors to cluster on the membrane.

The existence of a heterogeneous population of memory B cells with receptors of different antigen affinities also helps to explain why the antibodies generated after successive encounters with the antigen (called secondary, tertiary antibodies, etc.) are, in general, of higher affinity than the primary antibodies. A second encounter with the same antigen, which will generally be found in small quantities, will lead to preferential activation of the subpopulation of memory B cells with higher affinity receptors, since the subpopulation of memory B cells with lower affinity receptors will not be able to compete with their peers for antigen binding and will not be activated. This lower affinity population, in the absence of activation, will eventually disappear from the memory B cell population, so this will evolve to be composed of B cells with higher and higher affinity receptors. This evolution will take place because the process of competition and selection of the B cells with the highest affinity will be repeated at each subsequent encounter with the antigen, so as this happens a population of memory B cells will be generated that will have receptors for the antigen of increasingly higher affinity. This process is called **antibody affinity maturation**, and it is an important process in conferring increasingly strong protective immunity against a microorganism that is continuously present in the environment in which we live, and which we will therefore often encounter. This explains the convenience and need for booster doses of vaccines, that is due to the need for the vaccines to generate high affinity antibodies against the antigens.

The presence of memory cells also helps to explain a curious phenomenon called **original antigenic sin**. This phenomenon consists in the fact that the antibodies generated in posterior encounters with a microorganism that has mutated are directed against the antigens that these successive mutant microorganisms share with the initial one, but no antibodies are produced against new antigens generated by mutation of the original ones. To understand this better, let's look at an example. Let's suppose that the first time a child is infected with a flu virus, it has four antigens, A, B, C and D. If everything works normally, the child will generate four antibodies: one, anti-A; another, anti-B; another, anti-C and, finally, another anti-D. Three years later, the same child is re-infected with a flu virus. In this case, the virus has mutated and has only two antigens identical to those of the original virus and two new antigens, let us say that the virus now has the antigens A, C, E and F. Well, if everything works normally, the child will generate antibodies against the antigens present in the original virus, that is, anti-A and anti-C, but will not generate them against antigens E and F. Let us suppose, finally, that at the age of twenty years the same child, turned now into an adult, is infected again by a flu virus. In this case the virus has the antigens B, D, E and F. Well, the antibodies produced in this case will be anti-B and anti-D, but again no anti-E or anti-F antibodies will be generated.

The reason for this behavior lies in the generation of memory cells after the first encounter with the virus. These are the ones that will first find the antigen and react against it, helping its rapid elimination before other naïve B cells can find the new antigens present in the new virus strain. In addition to this process, the naïve B cells that could find the new antigen are not activated against it because in the presence of antibodies against the same antigen a cell signaling process is activated that interferes with the activating signal coming from the antigen. This prevents the generation of ineffective antibodies compared to those generated by memory cells. The generation of such ineffective antibodies would consume resources that are more efficiently devoted to the generation of antibodies of higher affinity from memory cells, which usually already generate antibodies of the appropriate class.

The greater ease with which memory T cells are activated in successive encounters with the antigen cannot, however, be explained by the fact that their antigen receptors have a greater affinity for the antigen. T cell antigen receptors do not undergo somatic hypermutation, so both naïve T cells and memory T cells possess receptors with identical affinity for the antigenic peptide:MHC complexes that they recognize. Studies on these cells have shown that memory T cells have modified their internal activation mechanisms after the stimulation of their receptors by antigenic peptide:MHC complexes. This modification facilitates the transmission of the biochemical signal from the membrane to the cell nucleus, which facilitates cell activation.

This may seem mysterious, but it is not when we consider that naïve T cells have safety mechanisms to avoid being activated too easily, which could cause high collateral damage and even lead to autoimmunity. These safety mechanisms include an enzyme located at the membrane that hinders the action of the other enzymes involved in signal transduction from the receptor to the cell nucleus. This enzyme, called **CD45**, can be produced in different variants (thanks, for those who want to know, to CD45 messenger RNA splicing). Naïve T cells produce the most restrictive version of this enzyme, while memory T cells produce a more permissive version that facilitates their activation.

That memory T cells are activated more easily than naïve T cells makes sense, since they have already passed all the safety barriers and were therefore completely activated by the antigen-presenting cells. Memory T cells, being derived from activated naïve cells, have also already shown their defensive utility, so it makes sense to allow their activation against a subsequent encounter with the same antigen to be easier.

Different T and B memory cells will accumulate in our body as we overcome infections throughout life. For this reason, towards the end of adolescence and the beginning of adulthood we will have a majority population of B and T memory cells that will be activated with some ease when their antigen is found. This is probably also the reason why the function of the thymus loses importance over time and in adulthood it is no longer necessary for the generation of new naïve T cells, since the initial repertoire of T cells has been activated and has generated memory cells against practically all antigens found in the environment where the organism lives. These cells are the ones that have already found an antigen that has been overcome, so the risks of a less restrictive activation are much lower than those we would run into if such an easy activation were not allowed. Interestingly, the immune system also estimates to some extent the likelihood of the risks it faces. A memory cell is much more likely to be activated against the antigen that its naïve precursor cell first encountered than against a self-antigen and generate autoimmunity. For this reason, the risk of allowing easy activation is less than the risk of not allowing it and thus making it easier for the microorganism to generate again the disease it generated the first time it infected us. The rapid activation of memory cells prevents the disease from developing by preventing the microorganism from establishing a site of infection. This is also the main advantage of vaccines.

In addition to the expression of proteins that facilitate activation upon a new encounter with the antigen, memory cells express other molecules that are necessary for their function. Some of these are expressed in common with effector cells, but all are particular to memory cells. Extensive studies over decades have identified many of these molecules, which also differ between different types of memory T cells. We won't get bored with an extensive list of them here, but we will mention some important examples. Here is the first: memory T cells express receptors for various chemokines that are important for directing them to lymphoid tissues or sites of infection, depending on whether they are central or effector memory cells. Another important molecule expressed by memory cells is **CD127**, which is part of the **IL-7** receptor, a cytokine that stimulates the growth and longevity of these cells. Finally, memory cells do not express, if not activated, effector cell-specific molecules, such as, for example, granzymes and perforin, in the case of memory CD8 T cells.

## 6.4.- Mechanism of action of vaccines

Once we have analyzed the generation and behavior of memory cells, we can begin to better understand the challenges of generating effective vaccines. These must, above all, generate an adequate population of memory cells, capable of defending us against the attack of the microorganisms against which we intend to vaccinate.

To achieve this goal, vaccines should ideally contain all the antigens presented by the microorganisms we wish to defend against, so that the greatest diversity of memory cells is generated against them. Ideally, the way in which the vaccine stimulates the immune system should also enhance the generation of memory cells instead of enhancing the generation of effector cells, whose generation is, in many cases, unnecessary, since the vaccine is rarely made with live attenuated microorganisms, i.e. microorganisms manipulated in the laboratory in such a way that they have lost virulence against humans, although they may be very virulent for other species.

Enhancement of memory cell generation instead of effector cells is possible and the factors involved are beginning to be known. These factors depend on the molecular conditions in which naïve T cells are activated by the antigen-presenting cells, which should come as no surprise. For example, it has been found that high levels of inflammation favor the generation of effector CD8 T cells, rather than memory cells, and in fact in cases of chronic infections where inflammation is high, memory CD8 T cells are not produced. High levels of inflammation are generated by cytokines and chemokines that facilitate the recruitment of effector cells to the site of infection. This increased inflammation also facilitates the transport of antigens to the lymph nodes and the transport of antigen-presenting cells to those organs, in addition to generating a molecular environment that stimulates naïve T cells probably in a different way than the stimulation that would occur under conditions of lower inflammation levels. In the latter case, it is true that there is an ongoing infection that needs to be overcome, but this is probably not as important as in the first case, allowing the immune system to generate more memory cells at the expense of effector cells without compromising the safety of the organism or the ability to eradicate the infection. It is fascinating how adaptive the immune system is in making the best decisions in the face of different types of threats and even in the face of different conditions in which the same threat, caused by the same microorganism, can occur.

The generation of an inflammatory state is fundamental to the effectiveness of the vaccine, as it needs to stimulate the cells of the innate immune system so that they can in turn stimulate the cells of the adaptive system in the right way to generate memory cells. This state of inflammation is generally achieved using **adjuvants**. We have not encountered them so far, but adjuvants are nothing mysterious. They are simply molecular components that activate antigen-presenting cells through the stimulation of their Toll-like receptors or other receptors capable of detecting microorganism-associated molecular patterns **(section 2.5.1)**. The need for adjuvants to stimulate an adaptive immune response was experimentally proven years before the discovery of dendritic cells. It was observed that purified antigens, such as proteins from viruses, do not usually trigger an immune response, i.e. they are not **immunogenic**. To obtain adaptive immune responses to purified antigens, it was found that it was essential to add dead bacteria or bacterial extracts to the antigen. This additional material was called adjuvant, as it helped to mount a response to the antigen (the Latin word *adjuvare* means 'to help'). Today, different types of adjuvants, not necessarily derived from bacterial components, can be used in vaccines, since they are necessary, at least in part, to activate dendritic cells so that they can present antigens in the absence of infection by living microorganisms.

However, in the early days of vaccination, this was done with living organisms. In fact, the invention of the vaccine, attributed to the English doctor Edward Jenner in 1796, would lead to the imprisonment for several years of the scientist who tried to repeat the procedure. Like so many and many inventions, Jenner's invention draws on previous achievements or knowledge. Jenner relied on the risky practice of inoculation and previous discoveries made by others. During the 18th century, smallpox was a terribly infectious disease that could cause mortality of up to 20% of those who contracted it. In an attempt to protect against it, a small amount of material taken from the pustules of the sick was inoculated under the skin of non-infected people. This practice had been carried out in China since at least the 10th century, and from there it had been imported to India, Turkey and Europe.

Inoculation of the suppurations resulted in a milder smallpox infection than the naturally occurring disease. People who got over this disease, if they got over it, were immunized. It is not clear why inoculation under the skin with human smallpox virus did not result in serious disease. One possible explanation is that the natural spread of the smallpox virus to the body does not occur through the skin, but through inhalation. The smallpox virus was adapted to infect the epithelial cells of the oral mucosa and pharynx and from these locations it could invade the local lymph nodes and reproduce rapidly. Twelve days after infection, the number of viruses had increased exponentially and viruses had invaded the blood, spleen, bone marrow and lymph nodes throughout the body. Under these conditions, the virus attacked the skin cells from the inside, generating the typical pustules of smallpox, which left deep marks if the disease was overcome.

However, it appears that if the initial infection was from an unnatural location for the virus, such as the epidermis, and not the oropharyngeal cavity, the virus was generally unable to invade the entire body from there. Obviously, we know this today, but neither the Chinese, nor the Indians, nor even Jenner knew the reasons. Among these we can consider several, such as, for example, a reduced capacity to quickly infect skin cells, which were infected more slowly than those in the mouth and pharynx, and whose infection occurred only when the virus had already reproduced at high levels in other parts of the organism. It is also possible that the inflammation generated by the inoculation, which in addition to the virus also introduced bacteria from the skin surface (I am afraid that the skin was not sterilized with alcohol, nor was the sharp object used for inoculation sterilized), stimulated antigenic presentation by the antigen-presenting cells that picked up the virus and also detected the bacteria. If this efficient antigen presentation occurred, it prevented the virus from having enough time to reproduce to a level that would allow it to invade the whole organism before the immune system had reacted and slowed down its progress. Here we are dealing again with the time factor as one of the most important factors in being able to defeat infections.

Based on the practice of inoculation, Jenner's idea was to inoculate not material from the pustules of human smallpox patients, which sometimes led to the development of a serious and fatal disease anyway, but from the pustules of people infected with bovine smallpox, which caused a less serious disease than human smallpox. In 1768, the also English physician John Fewster had realized that contracting the disease of bovine smallpox protected from contracting human smallpox. Today we know that this is possible because the bovine smallpox virus shares antigens with the human virus, which can stimulate an immune response that will therefore also be effective in neutralizing the human virus. At the same time, the bovine smallpox virus was not as virulent as the human virus, i.e. it did not cause such a serious disease, since it was not completely adapted to reproduce effectively in the human body. Today we also know that this is because, in general, viruses are optimally adapted to infect a given species, to which they cause more serious disease than other species, although they can also infect the latter. Again, this may be related to the speed of reproduction of the virus in one species or another, which affects the amount of virus that has been generated before the immune system has reacted to stop the infection. These different speeds of reproduction can be responsible for the generation of disease of varying degrees of severity in different species, and in different people. As we will see later, these factors were of enormous importance in generating the first vaccines in the laboratory.

John Fewster's observation that contracting bovine smallpox protected against human disease, combined with the practice of inoculation, gave Edward Jenner the idea of inoculating suppurations for bovine smallpox and checking whether this also protected against human smallpox. In 1796, in a trial that would be banned today on ethical grounds, Jenner inoculated an eight-year-old boy, James Phipps, the son of his gardener, with suppurations from pustules from the hand of a farmer infected with the bovine pox virus during milking of his cows. Seven days later, the boy suffered a fever and malaise which, fortunately for Jenner's garden, disappeared in a short time. A few days later, Jenner gave the boy several superficial punctures of the dreaded human smallpox virus. Fortunately, again, the boy did not contract the disease. To ensure that the procedure provided lasting protection, weeks later Jenner reinoculated the innocent child with material from human smallpox pustules. I do not know whether this inoculation was done in his garden or not, but the boy did not contract the disease this time either, which finally confirmed that the previous inoculation of the bovine virus could protect against human smallpox.

Apart from his scientific contribution, which is not lacking in luck, by modern standards of scientific research Jenner's experiment would today be a crime, both legal and scientific. First, it is exempt from the smallest particle of ethics that we are able to detect. Secondly, it is carried out only on one subject, not on a group of subjects, which is called the study population (although, given what we have seen, this is to be welcomed), in spite of which literally life or death conclusions are drawn based exclusively on the observation made with this subject. Of course, the experiment lacks a study control population, that is, for example, at least another child, perhaps the carpenter's son, who is inoculated with human pustules as a positive control (he should develop the disease) and at least one other child, perhaps the blacksmith's son, who is only inoculated with water or saline, as a negative control (he should not develop any disease). However, in those times of uncertain, dangerous biomedicine, Jenner's experiment led to the development of safe vaccines, first against the dreaded human smallpox, and then against other diseases.

Jenner was not the first to achieve immunization with the bovine pox virus, as a farmer in Dorset, southern England, had already successfully immunized his wife and two daughters in the same way, in 1774, during a smallpox epidemic outbreak at the time. However, it was not until Jenner conducted his experiment and published it, making it known to the medical and scientific community, that interest in the use and development of vaccines was sparked. By the way, we can now also understand why vaccines are called by this name, as the first vaccine was indeed of bovine origin (the latin word for cow is 'vacca').

## 6.5.- Types of vaccines

Following Jenner's successful vaccination, the medical and scientific community embarked on the development of other vaccines considered important in preventing serious diseases. As it seems logical, the first strategy used to generate vaccines was to try to obtain attenuated microorganisms, which, as we have said, are microorganisms whose virulence has decreased to the point of not generating disease in humans, although they can generate an immune system response capable of protecting the body against the original microorganism that does cause disease. Vaccines composed of attenuated organisms are, of course, called **attenuated vaccines**.

Attenuated vaccines are used in general for immunization against viral diseases and are generated by a targeted evolution of the original microorganism. This evolution is induced by growing the microorganism in cells or animals of another species. Normally, a particular pathogenic microorganism is adapted to infect one or a few related species. This adaptation is the result of its genes producing certain proteins that fit well with the proteins of the host organism and allow efficient infection.

Viruses need to bind to some protein on the surface of the host cells or otherwise they cannot penetrate them, where they irretrievably need to reproduce. The host cell protein functions as a lock, while the virus protein functions as the key that fits that lock and allows it to open it and thus enter the cells. Just as the key to our outer door might fit into our neighbor's lock but does not allow his or her door to be opened, so do the key and lock proteins of viruses and cells from neighboring species. Viruses that infect animals cannot generally infect humans unless they have a molecule that functions as a sort of master key that allows several doors to be opened at once, or the locks on the cells of human and other host species are identical or very similar. Even in this case, however, the virus will infect some species more easily than others, because in the case of viruses, it is not just a question of opening the door, but of using efficiently the resources of the cell of the species in question once the virus has introduced its genetic material into it. A virus adapted to one species, even though it is capable of entering the cells of other species and infecting them, will not reproduce exactly with the same efficiency in both types of cells and will therefore be more virulent in one species than in the other.

However, most viruses are microorganisms with a high capacity for mutation. Among the millions of viral particles generated in an infection, mutants have always been generated that are less effective than the initial ones in infecting the cells of the original species, but that, by chance, have modified their key so that it can fit well in the molecular lock of another related species. These mutant viruses can be grown in the laboratory in cells of that species for several generations and thus allow them to evolve by mutation and selection and eventually generate viruses whose key and other proteins have adapted perfectly to the lock of the new species and to the internal machinery of its cells, but which will now not fit well into the human species, which it will therefore have difficulty infecting. The viruses generated in this way are attenuated viruses and can be used as a vaccine.

Attenuated vaccines have certain advantages over other types of vaccines that we will discuss later. First, they properly activate the immune system so that it generates adequate protection. For example, they safely induce the generation of the most effective antibody classes against the virus in question. In addition, they provide long-lasting immunity that requires fewer booster doses, since the microorganism is alive and, although slow, reproduces and infects cells before it is completely eliminated by the immune system. In other words, attenuated vaccines generate a real, albeit mild, infection that the immune system must overcome. The vaccination process is therefore the closest thing to a natural infection that can be aspired to without generating the disease. On the other hand, attenuated vaccines are generated at low cost. Finally, as we will see in more detail later, some attenuated vaccines generate benefits that go beyond protection against the microorganism for which they are vaccinated.

However, attenuated vaccines also present a higher risk than vaccines composed of dead organisms or molecular components of these organisms. The reason for the higher risk of attenuated vaccines lies in their potential ability to adapt back to the human species. Although the microorganisms in attenuated vaccines have several mutations, not just one, that differentiate them from the original microorganism, because attenuated viruses are alive and will reproduce in human cells, even if they do so at a slower rate and with less efficiency than the original virus, the activity of the immune system stimulated by the vaccine must be able to eliminate the infection before it generates appreciable symptoms of disease. This is normally the case. However, it is always possible, though unlikely, that by bad luck some attenuated virus will mutate early during the vaccine-induced infection so that this mutation will allow it to effectively re-infect human cells and cause the disease it was intended to protect against.

Of course, the likelihood of this reversal depends on the time it takes for the immune system to clear the infection, since as long as live attenuated viruses are present, they can reproduce and will always be able to mutate. Fortunately, attenuated vaccines usually induce a rapid immune system reaction, so attenuated viruses injected with the vaccine are quickly eliminated in normal people. However, if an attenuated vaccine is given to an immunodeficient person, usually a child, he or she will not be able to clear the infection. The attenuated virus will continue to reproduce, since defects in the immune system of the immunocompromised child will not allow it to be eliminated. Under these conditions, it is only a matter of time before the virus readapts to human cells and causes a potentially fatal disease, since the immune system will not be able to fight it off effectively. This problem can arise with the attenuated polio vaccine, a disease caused by a virus that attacks many cells in the body and, in the most severe cases, can even attack motor neurons, causing paralysis. Infants who are not known to be immunodeficient in producing immunoglobulins, when vaccinated with live attenuated polio virus, are at much greater risk of reversion of the vaccine to a virulent strain of the virus causing disease. This happens because, in immunodeficient children, the attenuated vaccine cannot generate antibodies that would remove the virus from the gut (the organ first infected by this virus before infecting others from there). Since it is not eliminated by specific antibodies, the virus continues to reproduce and can therefore mutate. In some cases, the mutations can reconstitute the normal phenotype of the virus, which will allow it to attack the neurons and cause a fatal paralytic disease. In immunodeficient children, unable to eliminate the attenuated virus, the appearance of this virulent mutation is only a matter of time, so these children, if vaccinated, will sooner or later develop polio.

This also illustrates the fact that most antiviral vaccines, to be effective, must be able to generate virus-neutralizing antibodies, i.e. antibodies that bind to the virus and prevent it from attaching itself to the molecular lock it must open to penetrate the cells and infect them. The antibodies must be of the right kind and be abundantly located in the tissues where the viruses can penetrate the body. In the case of polio virus, which is transmitted orally, it is essential that the vaccine induces the generation of IgA class antibodies, which are the only antibodies that can be transported from the outside of the intestinal tract, where they are produced in the lymphoid tissues of the intestine, to the inside of the intestinal tract, where they exert their protective function by binding to the virus and preventing it from entering the body through the intestinal wall. In contrast, viruses neutralized by antibodies cannot pass through the intestine and are excreted in the stool.

We see here that sometimes the effective protective mechanisms that must be induced by vaccines are not the same as those put in place to eradicate the infection. In the case of polio, the primary infection occurs because the body does not have specific IgA antibodies against the virus, obviously, since the infection has not happened before. In the absence of this antibody, the infection occurs and must be defeated not by the generation of IgA, but by the generation of another class of neutralizing immunoglobulins, IgG, which act in the body's fluids and not on the surface of the intestine, as well as by the activation of cytotoxic CD8 T cells, which are the main cells involved in the fight against the virus.

The generation of antibodies of the right kind depends on the initial route of entry of the virus and the dendritic cells it first encounters and which will present its antigens to the T cells. Therefore, the route of administration of the vaccines is important. For the generation of neutralizing antibodies of the IgA class, in addition to IgG, the preferred route of entry is the oral route, and the attenuated polio vaccine is administered this way. However, due to the risks of this vaccine, another vaccine has been developed that is composed of inactivated, i.e. dead, viruses that are unable to infect. This vaccine is administered by intramuscular injection. Although this type of vaccine induces only IgG generation and little or no IgA, if the IgG levels produced are high enough, the vaccine is very effective and very safe.

This leads us to mention that, in addition to attenuated vaccines, **inactivated vaccines** have been developed, which constitute a different class of vaccines. These include one of the first vaccines produced, the rabies vaccine, and also one of the two types of seasonal flu vaccines available, which should be given every year to people at high risk of contracting this disease and suffering serious consequences (the other type of flu vaccine is an attenuated vaccine). The inactivation of microorganisms is achieved by physical means, such as heating or irradiation, or by chemical means, such as treatment with formaldehyde, a substance that binds to proteins and denatures them, thus preventing them from working. Inactivated vaccines are safer than attenuated vaccines in that they cannot cause disease by reverting to the original microorganism by mutation. However, the fact that they do not produce a mild infection to be overcome results in less effective subsequent protection than that generated by attenuated vaccines. To counteract this problem, booster injections are often given, which is not always easy in some parts of the world and is more stressful for children and parents.

Another type of vaccine is one that generates neutralizing antibodies against some bacterial toxins. These include diphtheria and tetanus vaccines. We have already seen at the beginning that toxins are proteins, secreted by bacteria, capable, like viruses, of binding to a protein on the surface of the cells and, thanks to this, penetrating into the cells and killing them, thus releasing nutrients that are necessary for the microorganism. Without the activity of the toxin, the microorganism cannot reproduce at high speed, so the immune system has more time to set in motion its defense mechanisms and eradicate it, before it generates any appreciable disease. Under these conditions, vaccines that induce the production of antibodies that bind to the toxins and through that binding neutralize them and prevent them from penetrating into the cells will also provide effective protection against the disease in question. In these cases, it will therefore not be necessary to generate attenuated or inactivated vaccines with the whole microorganism; it will be sufficient to induce immunity to the toxin itself to prevent it from causing disease if the bacterium attacks us, since, with the toxin neutralized, the bacterium will in fact behave like an attenuated organism.

Of course, we cannot simply vaccinate by injection with the toxin isolated from the bacteria, as it is toxic. Vaccination must be done with the inactivated toxin, also by physical means, such as heating, or chemical means, such as treatment with formaldehyde. Inactivated toxins are called toxoids, and vaccines formed from them are called **toxoid vaccines**. In this case, it is essential that the inactivation does not destroy all the toxin natural epitopes, since the antibodies that the toxoid generates must be able to bind to those epitopes to neutralize the toxin.

Fortunately, bacterial toxins are generally made up of two protein subunits. One of the subunits plays the role of a key to bind to the cell membrane protein that acts as a lock. The other subunit of the toxin is usually an enzyme and is the one that performs the toxic function once inside the cell, catalyzing a chemical reaction that disables one of the cell vital components, or even prevents nerve transmission affecting the life of the whole organism, as is the case with tetanus toxin. This implies that, if the subunit capable of binding to the protein lock of the cells is neutralized, the toxins cannot enter its interior and cannot exert their toxic function. This also implies that the neutralizing antibodies generated against the key subunit should be enough to neutralize the activity of the entire toxin by preventing its entry into the cells, and usually are.

Consequently, many toxoid vaccines are formed exclusively by the subunit of the toxin that binds to the cell lock, in its natural state, i.e. keeping intact the epitopes to which neutralizing antibodies can bind to neutralize the natural toxin in case of infection. The subunit that functions as a key is not toxic by itself, so the administration of these vaccines is absolutely safe.

Toxoid vaccines formed with only one of the subunits of a bacterial toxin naturally lead us to talk about **subunit or component vaccines**. In this type of vaccine, the subunits come not from the toxins but from the very microorganisms that are intended to be vaccinated against. To generate this type of vaccine, one or more components of the microorganism that are the main targets of the immune system activity must be isolated and purified. For example, if it is necessary for the vaccine, to be effective, to generate neutralizing antibodies, the components of the vaccine must be present on the surface of the microorganism so that the antibodies generated can bind to them and thus block their ability to infect, reproduce and generate disease. The pertussis vaccine is an example of a vaccine consisting of components of the microorganism that cause these diseases.

Another type of vaccine is the **conjugate vaccine**. This is the most interesting type of vaccine, because it manages to manipulate the immune response to make it more effective than the original response to the microorganism. This vaccination is only possible in the case of specific diseases that can be experienced, in general, by young children. To understand why and how this vaccine works, we must go a little deeper into the fascinating mechanisms of the immune system.

The first thing we must know is that there are two types of antigens capable of generating a humoral immune response, that is, the one constituted by the generation of antibodies. These two types of antigens are called **T-dependent and T-independent antigens**.

T-dependent antigens, as their name suggests, depend on the action of T cells so that antibodies against them can be generated. These are the type of antigens that we have studied in this book so far. Remember that B lymphocytes, to carry out somatic hypermutation and immunoglobulin class switching, without which they cannot generate effective antibodies, need to receive cytokines produced by helper CD4 T cells. These cells will not produce and secrete cytokines unless, firstly, they have been activated by antigen-presenting cells that present antigenic peptides in their MHC-2. In addition, the T cells must be in turn stimulated by B cells through the presentation, in their MHC-2, of the same peptides that activated them, derived from the foreign microorganism that B cells have had to capture by binding to it through their antigen receptors, and had to internalize, digest and process it to bind peptides to MCH-2 molecules. Thus, the proteins of the microorganisms, in general, are T-dependent antigens, since they must be degraded to peptides that must be presented by MHC-2 molecules to allow the recruitment of helper CD4 T cells.

However, not all components of a microorganism are proteins. In particular, many bacteria are surrounded by components that are carbohydrates or carbohydrates bound to lipids. These components form a capsule that surrounds the bacterium and makes it resistant to phagocytosis by neutrophils, macrophages and dendritic cells. This class of bacteria with protective capsules are called **encapsulated bacteria** and cannot be directly phagocyted. In the absence of phagocytosis, the bacteria cannot be captured, digested and their peptides presented in the MHC-2 of the antigen-presenting cells, so these cells cannot activate specific CD4 T cells that would be needed to help the B cells that would have been able to detect them through their receptors and, this time, incorporate them thanks to them, digest them and present their peptides in the MHC-2, to generate antibodies. For antigen-presenting cells to capture them, antibodies would be required to bind to the surface of the bacterial capsule and thus opsonize the bacteria, but these antibodies cannot be produced unless the bacteria are captured. As we can see, we are in a kind of vicious circle and during their evolution some bacteria have managed to take advantage of this apparent weakness of the immune system.

Fortunately, some B cells do not always need the help of T cells to generate protective antibodies. T-independent antigens, as their name also suggests, can induce antibody production by some mature B cells without the help of CD4 T cells.

There are two types of T-independent antigens (TI), also called thymus-independent (since T cells are produced in the thymus) named **TI-1** and **TI-2**. The difference between the two refers mostly to the way in which they can activate B cells without the help of T cells. TI-1 antigens possess a chemical nature that allows them to interact with B-cell Toll-like receptors and send a signal through them that stimulates cell activation and reproduction, i.e. mitosis. They are, for that reason, B-cell mitogens.

Once stimulated, the B cells will differentiate and proliferate and become antibody-producing cells, but since the antigen has not bound and has not stimulated its antigen receptor, but only the Toll-like receptors, the antibody they will secrete will not bind to it. This is a drawback of the TI-1 antigens, which can stimulate a huge number of B cells to mature and generate antibodies that, in principle, will not be of any use.

However, the above effect is only observed when the amount of TI-1 antigen is very high, i.e. under laboratory conditions that do not represent what actually happens. In reality, when bacteria initiate an infection they do not do so in large numbers, so those that have molecules in their structure that act as TI-1 antigens are not sufficient to stimulate B cells, unless these, at the same time, react through their antigen receptors with another epitope of the same bacteria. In this case, the B cell receives two stimulating signals simultaneously: one, weak, through the Toll-like receptors, and one through the antigen receptors. These two signals combined are sufficient to stimulate the B cells that can receive them. Since only B cells that possess an antigen receptor capable of interacting with a sufficient affinity with some component of the bacterium will be stimulated by the TI-1 antigens, these cells will produce antibodies of the IgM class that will be able to bind to the bacterium and stimulate thee complement system to promote opsonization and phagocytosis.

The nature of TI-2 antigens is different. They do not act as indiscriminate mitogens, but rather activate B cells due to their repetitive nature, which is capable of binding numerous B cell receptors on the membrane. This assembly and intercrossing of receptors on the membrane allows the stimulation of intracellular mechanisms that generate a very strong activating signal, capable of making only the B cells that have detected them through their receptors differentiate and reproduce. This leads to the production of IgM antibodies by the B cells, although they can also produce IgG, specific to the bacterial antigen that has stimulated the B cells.

The fact that the immune system has the ability to generate antibodies independently of the T cells allows for a very rapid production of these, because it does not need to wait until the T cells have been activated, a process that takes several days. This rapid initial production of antibodies, even if they are of the IgM class and of low affinity, in general, can be important to begin to contain the progress of the infection and thus give time for other B cells to receive the help of T cells later and undergo the somatic hypermutation and class switch that will allow the total eradication of the infection.

The above brief analysis allows us now to better understand the importance of conjugate vaccines. These vaccines, combine, conjugate, two different antigens. In this sense, are also multiple combined vaccines and can be used in some cases to vaccinate against several diseases at the same time in a single injection, as is the case with other combination vaccines, such as the triple anti-viral vaccine against measles, mumps and rubella (MMR vaccine), composed in this case of attenuated viruses. However, combination vaccines are not necessarily conjugate vaccines. The uniqueness of conjugate vaccines is that they not only combine two different antigenic components of two different pathogenic microorganisms, but do so in such a way that both components are physically linked to each other by irreversible covalent bonds, , i.e. they are linked in such a way that they cannot be separated except by enzymatic degradation of these components inside the antigen-presenting cells or B lymphocytes. In addition, one of the linked components is an antigenic protein isolated from a pathogen, while the other is a non-protein component in nature, usually a carbohydrate isolated from the outer polysaccharide capsules of pathogenic encapsulated bacteria. This latter component alone would not be able to generate antibodies in a T-dependent manner, i.e. it is a thymus-independent antigen.

The physical binding of these components gives conjugated vaccines an important property. This is to be able to transform the thymus-independent antigen into a thymus-dependent one. This is because the thymus-independent antigen is physically bound to a thymus-dependent one. Let's see how this changes its nature.

After injection of the vaccine, dendritic cells will capture the antigen, activate and begin to degrade it, leading to the generation of peptides that can be presented in MHC-2 molecules for activation of naïve CD4 T cells, which will be activated to effector helper T cells. At the same time, after the injection of the vaccine, several different B cells will also be able to detect and capture the antigen by its different epitopes. All these cells will receive a stimulating signal and will start to be activated. Moreover, all of them will internalize the antigen linked to its receptors and will be able to degrade it inside, generating peptides from the protein component, peptides that will be presented by the B cells in MHC-2 molecules. This peptide presentation will enable these cells to receive the collaboration of helper CD4 T cells that have been activated by the same peptides bound to MHC-2 molecules presented by the dendritic cells. This collaboration will allow the activated B cells to undergo somatic hypermutation and class switching, and generate more effective antibodies against both the protein component, (for which this process would have occurred even if it had been injected alone) and, importantly, the non-protein component, which is a thymus-independent antigen that would only have generated low affinity IgMs if it had been injected without being conjugated to a protein, but which can now generate high affinity IgG or IgA antibodies.

The importance of transformation from a thymus-independent antigen to a thymus-dependent antigen is that young children do not yet have a fully mature immune system and cannot effectively produce antibodies against thymus-independent antigens, but they can do so against thymus-dependent antigens. Defense against encapsulated bacteria, such as *Streptococcus pneumoniae*, also called pneumococcus, requires the production of neutralizing and opsonizing antibodies against the polysaccharide coating of these bacteria, which young children cannot produce, since this is a thymus-independent antigen. However, the conjugate vaccine will allow young children to produce polysaccharide antibodies in a thymus-dependent way, which they can do normally despite their young age. This gives vaccinated children protection against *Streptococcus pneumoniae* that could not be achieved by injecting any combination of bacterial antigens not physically conjugated by chemical means, or even by injecting attenuated bacteria.

Pneumococcal conjugate vaccine consists of the chemical bonding of the bacterium capsule polysaccharide with the toxoid derived from diphtheria toxin. This combination generates immunization against both diseases, through the production of memory B cells that will be activated in subsequent encounters and will generate IgGs or IgAs antibodies of high affinity. In these conditions, the eventual encounter with the bacteria will not cause disease and, on the contrary, will reinforce the presence of the memory cells.

However, the streptococcus that causes pneumonia is a bacterial species that protects itself in 91 different ways. In other words, its outer capsule can be one of 91 different forms. Antibodies to one of these forms do not protect against the others, so vaccines against this bacterium must be composed of the conjugation of diphtheria toxoid with at least the most predominant forms of the polysaccharides in the outer capsule to induce protection against the more frequent streptococcal variants. The vaccine was thus generated with seven different polysaccharides isolated from the seven predominant bacterial variants. This prevents the infection of children with these variants, which are being eradicated from the bacterial population, but has favored that other less predominant variants are increasing their frequency of infection. This may force a change in the composition of the vaccine in the future, so that it contains the polysaccharide variants that at any given time carry a higher risk of infection.

The last type of vaccine we will explore briefly are **vaccines made up of nucleic acids**. These include **DNA vaccines** and **RNA vaccines**. Obviously, it is not the case that the vaccine generates antibodies or memory T cells against the nucleic acids which, by the way, if they are pure and lack associated proteins, would also be thymus-independent antigens. In this case, the intention is that the genetic information contained in the nucleic acid is translated into proteins and that these will induce an immune response against them. It is therefore a question of the DNA or RNA containing the genetic information to produce proteins specific to the microorganism against which the vaccine is intended. These foreign proteins would be the ones that would induce the immune response, helped by appropriate adjuvants injected at the same time as the nucleic acid. DNA vaccines are still in the process of research and development, and there is no DNA vaccine approved for use in humans, although there are for veterinary use.

DNA vaccines have some interesting advantages from an immunological point of view, since they can be manufactured in such a way that the proteins induce preferably cellular or humoral responses. They are also relatively easy to produce and are very stable and hold up well without degrading at room temperature for a long time, making them suitable for preservation and for transport to various regions of the world where they may be needed.

Of course, DNA vaccines have disadvantages as well. The main one is that the DNA injected with the vaccine could be integrated at some random site on a chromosome and modify the genome of the cells that incorporate the DNA. This incorporation must necessarily occur for the cells to produce RNA first and then proteins encoded in the DNA. If the integration of the DNA takes place at a harmless site, i.e. one whose modification does not cause particular problems for the cells, there would be no problem, but, from time to time, in one or another cell of one or another person, the DNA may integrate at a site on some chromosome which, when modified by the integration of foreign DNA, may generate a serious cellular problem. This could be, for example, a change in the cell genome that transforms or helps to transform normal cells into tumor cells. If this happens, it could eventually, several years after vaccination, cause the generation of a tumor. For this reason, DNA vaccines are unlikely to be used in the human case unless the benefits of DNA vaccination far outweigh the risks and there is no alternative vaccine of another type.

RNA vaccines circumvent this problem. In this case, the genetic information is contained directly in the RNA, and it is not necessary for the RNA to be generated from the genetic information contained in the DNA. The vaccine RNA would penetrate the cells and be directly translated into the immunogenic proteins that would induce the protective immune response, i.e. the vaccine RNA would function as a cellular messenger RNA. The RNA would not be integrated into the genome in any case, nor would it modify it in any way, so it would not generate potentially dangerous mutations. RNA can also be designed to increase its stability within cells and its speed of translation, thus generating a high level of expression of immunogenic proteins. However, the stability of RNA vaccines is much lower than that of DNA vaccines, and they could easily be degraded and inactivated before being administered if precautions were not taken to avoid this. RNA is easily attacked by enzymes called RNAases, which are often found in the environment, and present in fluids such as saliva. Any microscopic droplets of saliva that might come into contact with the vaccine preparation could inactivate it completely. In addition, the resistance of RNA to exposure to relatively high temperatures, such as summer temperatures, is much lower than that of DNA. On the other hand, RNA vaccines can stimulate the production of type I interferons against the vaccine's own foreign RNA component, an unwanted immune response that can compromise or affect the overall efficacy and safety of RNA vaccines.

## 6.6.- Adjuvants

Modern techniques in Molecular Biology allow the cloning of viral or bacterial genes, their use in the laboratory and the production of the most suitable antigenic proteins for the generation of vaccines. These vaccines, derived from the artificial recombination of genes and DNA fragments, are called **recombinant vaccines** and are currently in use. They are component-type vaccines that are absolutely safe.

A problem with these component vaccines produced and purified in the laboratory, however, is that they lack those microorganisms-associated molecular patterns **(section 2.5.1)** necessary to stimulate dendritic cells to present antigens in the lymph nodes. In the absence of these patterns, the injected antigens, instead of inducing an immune response against them, induce their tolerance, i.e. not only will they not protect against the microorganisms, but may even favor their infection by inhibiting, rather than activating, an immune response against them. This is because in the absence of the cytokines generated by antigen-presenting cells in the process of inflammation, mature naïve B and T cells interpret that the antigen they now encounter, since it has not activated the cells of the innate immune system, must be a self-antigen that they have not detected at the time of their development. Remember that, under these conditions, these cells enter a state, called anergy, **(section 2.5.4.1)**, in which they can no longer be activated in response to that antigen.

For this reason, vaccines, in addition to antigens, must contain **adjuvants** which, as we have already mentioned, are substances that help stimulate the immune system, especially the innate immune system. These substances are not immunogenic in themselves and in the absence of the antigen specific to the microorganism against which it is desired to vaccinate, they do not generate protective immunity or memory cells.

Numerous substances or combinations of substances have been used as adjuvants. One of the best known in research is **Freund's adjuvant**, used to stimulate the immune response of laboratory animals. Its use in humans is prohibited, due to its toxicity. It couldn't be other way, because it is formed by a mixture of dead and dry tuberculosis mycobacteria and mineral oil. Currently, recommendations are followed to limit its use even in laboratory animals, as it can cause necrosis at the injection site. Freund's adjuvant stimulates a strong inflammatory reaction and a high production of TNF-$\alpha$, a cytokine which, by acting on the endothelium, relaxes it, and promotes coagulation to prevent the spread of infectious microorganisms, hinders blood circulation and oxygen supply. A normal production of TNF-$\alpha$ is beneficial to defend us from an infection, but a too high production can cause a serious deficiency of oxygen supply to the area and, therefore, cause necrosis of cells so, under these conditions, it is too toxic and harmful. The reason why Freund's adjuvant stimulates such a high production of TNF-α is probably because it contains many more mycobacteria than normally enter the body with an infection, resulting in the stimulation of more cells of the innate immune system than are normally stimulated under normal conditions.

The pathogens that attempt to infect the organism possess a diversity of associated molecular patterns that communicate the appropriate information to the cells of the innate immune system, so that they are activated correctly and allow the most appropriate decision-making and defense mechanisms to be put in place. The different molecular patterns associated with the same microorganism generate a synergy among them, by activating at the same time different Toll-like receptors and other types receptors, typical of the cells of the innate immune system. This is the reason why attenuated vaccines do not need strong adjuvants, since the attenuated microorganism itself is already capable of adequately activating the innate immune system. This will allow the presentation of antigens to CD4 T cells to allow that these can activate CD8 T cells correctly, if necessary, as well as activate B cells to produce the correct class of antibodies against the microorganism that we wish to fight. However, other types of vaccines that are not based on attenuated microorganisms do need adjuvants that adequately stimulate the innate immune system.

Traditionally, the adjuvants used have been specific substances, such as aluminium hydroxide, mineral oils or even saponins extracted from plants. Except for Freund's complete adjuvant, which contains dead mycobacteria (the incomplete one is only mineral oil) and cannot be used in humans, adjuvants do not contain microorganism-associated molecular patterns. Fortunately, recent research is advancing in the formulation of adjuvants that generate a synergy similar to that naturally generated by the molecular patterns of live or attenuated microorganisms. This synergy is hoped to achieve optimal activation of the innate immune system, which in turn would lead to optimal activation of the adaptive immune system and increased generation of memory cells to provide protection against the microorganism against which vaccination is intended. Thus, generating a good vaccine requires not only that the antigen selected to produce it is suitable, but also that the adjuvant molecules used are suitable. Even though vaccines have been known for more than two hundred years, there is still much needed research and knowledge to generate safer and cheaper vaccines, as well as vaccines against diseases for which effective vaccines are still lacking. Currently (year 2020, as I write this) we have only twenty-six effective vaccines, listed on the World Health Organization website. Twenty-five vaccines for important diseases are being investigated. These include diseases of the incidence and importance of malaria and AIDS, and of course, COVID-19. Hopefully, research will soon be successful and effective vaccines for at least the deadliest diseases will soon become a reality.

## 6.7.- Secondary benefits of vaccines

Unfortunately, the anti-vaccine movements have focused on the harmful effects that vaccines can sometimes produce. Because these effects occur, they have sometimes been cited by anti-vaccine activists as evidence supporting that vaccines can be harmful. We have already seen, for example, that attenuated vaccines, which are the first vaccines to be produced, could cause serious disease in immunosuppressed people. However, most of the harmful effects attributed to vaccines are false, such as vaccines causing autism. The origin of this myth comes from the publication, in 1998, of a study performed with only 12 children. The study could never be confirmed with a larger number of children and was later withdrawn, i.e. considered to be false, as if it had never been published, but the publicity given to it has lasted until today. Let us consider that around 120 million children are vaccinated every year. If vaccines were to cause autism or other problems, we would know by now without the need for further studies. To add evidence to the grievance, it is known that the study was funded by an attorney who had an interest in showing that autism was a harmful effect of vaccines. His intention was to sue the pharmaceutical companies for millions of dollars on behalf of those affected by autism who had been vaccinated. This myth, thanks to the anti-vaccine movements, has caused much illness, pain and even death to people who have stopped vaccinating themselves and their children.

Media attention to disasters, scandals, and various problems is, in general, much greater than that devoted to beneficial aspects of reality. The positive aspects, in my opinion, are less credible than the negative ones, and perhaps that is why the media lose audience if they spend too much time on good news. For this reason, a single study that wrongly indicated the existence of a problem caused by vaccines was amplified in an absolutely irrational and disproportionate manner. Interestingly, the same is not true of studies that indicate that, far from causing problems, vaccines generate health benefits not only related to protecting us from the microorganism against which we are vaccinated.

Observations that vaccines can generate benefits not only related to the disease they protect against were made shortly after the vaccines were first used. In fact, few years after the smallpox vaccine developed by Jenner was introduced, it was observed that it not only protected against smallpox, but also against other diseases such as measles, scarlet fever, syphilis and other problems related to the functioning of the immune system, including the development of allergies. Furthermore, following the introduction in Sweden, in the early 20th century, of the tuberculosis vaccine, the bacillus Calmette-Guerin (BCG), it was found that overall mortality was almost three times lower among vaccinated young children. This significant decrease in early childhood mortality could not be explained by the lives saved by the vaccine alone, as tuberculosis caused mortality in older children. Therefore, the conclusion of this finding was that the BCG vaccine causes non-specific immunity, a conclusion that cannot be avoided considering that, in those years, children did not die practically for reasons other than infectious diseases.

These observations have been confirmed over the decades of BCG vaccine use, and more recently, the benefits and potential reasons for them have been investigated. Recall that the tuberculosis vaccine was invented by French physician Albert Calmette and his assistant Camille Guérin in the early 20th century. This vaccine is an attenuated vaccine, made up of live bacilli that cause bovine tuberculosis. These bacilli are as virulent as those causing human tuberculosis and cause severe tuberculosis in both humans and cattle. However, Calmette and Guérin set out to culture the bacillus in the laboratory under various conditions in an attempt to obtain attenuated strains that could be used as vaccines. They found that the bacillus could be grown in a kind of potato-glycerin-based soup, in which it lost some of its virulence. After repeatedly culturing the bacillus 239 times and analyzing its virulent properties for 13 years, progressively selecting the less virulent bacilli, Calmette and Guérin obtained the variety of bacillus that bears their name (known by the initial letters: BCG), which is much less virulent than the original, and has been used as a vaccine against tuberculosis in hundreds of millions of people in the world.

The effectiveness of this vaccine against meningitis tuberculosis is very high, but it is not as effective against pulmonary tuberculosis. However, despite being an attenuated vaccine that carries greater risks than modern vaccines, generated from inert molecular components of the microorganisms, the BCG vaccine has been continuously used since 1921 and is the oldest vaccine still in use, in the absence of a more effective one.

Perhaps because of the higher risk initially associated with this vaccine, as well as its surprising beneficial effect on child mortality, the scientific community has been interested in studying more closely the health status of people vaccinated with the BCG bacillus. The findings reveal new and unsuspected health benefits of this vaccine.

Firstly, the impressive fact that people who had suffered from tuberculosis had a lower incidence of cancer was noted. This led to the discovery that the BCG vaccine is quite effective in fighting bladder cancer and today more than three million bladder cancer patients have been successfully treated with direct injections of this vaccine into the bladder, which have been used since 1977. This vaccine is therefore one of the first cancer immunotherapy strategies that have been used. The mechanisms by which it works are not yet known with certainty. It is believed, however, that the fact that the vaccine contains a living organism stimulates the immune system in ways that vaccines that are only made up of inert molecules cannot.

The BCG vaccine also produces benefits against allergic and autoimmune diseases. Those who receive the BCG vaccine are protected from developing asthma and other allergic diseases as well. The reason for this effect is better understood and is related to the development of the immune system from childhood, which can be adversely affected in environments too devoid of microorganisms. This effect is called the "hygiene hypothesis", which maintains that too much hygiene disorients the immune system, by eliminating too many enemies of the environment to fight against, which encourages the immune system to react to substances it should tolerate. The BCG vaccine, because it consists of a living microorganism, would correctly "educate" the developing immune system and prevent it from investing unnecessary energy in fighting non-existent enemies.

A similar protective effect of the BCG vaccine has been observed against autoimmune diseases, where the immune system makes the serious mistake of attacking the body itself, which it should, however, protect. Such diseases include some of the seriousness of multiple sclerosis and type 1 diabetes mellitus, a disease caused by the immune system attack on the cells of the pancreas that produce insulin. The elimination by the immune system of these cells leads to the inability to make insulin and, consequently, to diabetes. The BCG vaccine decreases the risk of developing these and other autoimmune diseases.

Another vaccine that has important benefits is the measles virus vaccine, which is part of the measles- mumps-rubella (MMR) vaccine. From 2000 to 2017, measles vaccination has reduced measles deaths by 80 per cent. However, in the absence of vaccination, transmission of the measles virus is highly likely. Ninety percent of unvaccinated people will be infected with the virus if they come into contact with another infected person. Transmission is easy, as the measles virus is spread through the air from aerosols produced by coughing or sneezing. These are the reasons why the anti-vaccine movement has led to a 300% worldwide upsurge in the disease, now affecting more than seven million children and directly killing more than 100,000 each year.

Measles is a disease that lacks specific treatment and, very importantly, causes immunosuppression, i.e. it leaves the body defenses very weak. This allows for the attack of other infectious microorganisms that can cause serious illness and even death, including intestinal infections that cause diarrhea that is very difficult to control, and pneumonia.

Historically, the first evidence found for the immunosuppressive effect of the measles virus was that children who had overcome the disease stopped responding to the tuberculin test. Those of a certain age may still remember this test, given to assess whether people had been infected by this mycobacterium or determine whether the BCG vaccine against tuberculosis had been effective. A positive test indicated one thing or the other. Well, a positive result on this test could progress to a negative test result after suffering from measles, indicating that the disease seemed to make the immune system forget that it had been vaccinated against tuberculosis, and also suggesting that measles could probably make the immune system forget about other vaccines, as well as diminishing the natural immunity developed against microorganisms that children's immune systems have encountered throughout their development. Subsequent studies showed that suffering from measles disease increases the likelihood of contracting other infectious diseases, and the mortality associated with them, up to five years after the disease is overcome. The data also indicated that measles may be associated with 50% of infant mortality caused by other infectious diseases.

However, because vaccination, which has been widespread until now, has greatly reduced the number of cases of the disease, it was not thought necessary to investigate the degree of immunosuppression caused by the measles virus, nor the cellular or molecular processes by which it causes it. In 2019, an international group of European and American researchers wanted to start remedying this sad situation. The researchers studied the effects of measles on the immune system with the most modern tools available. Using the VirScan assay, a technique that identifies all the antibodies in the blood against pathogenic microorganisms, the scientists studied 77 unvaccinated children in the Netherlands before and after a natural measles infection. The results of this study indicated that measles caused a decrease in the repertoire of protective antibodies against other microorganisms in a range that varied from 11% of this repertoire in the most fortunate children to 73% of the repertoire in the most affected. The children's immune system seemed to have forgotten that it had fought many microorganisms in the past. This phenomenon has been called **immune amnesia**.

The researchers also studied whether the decrease in protective antibody repertoire occurred equally in children vaccinated against measles. It was reasonable to think that perhaps the vaccine, which mimics a measles virus infection, would cause similar effects. This is not what happened. While the vaccine was effective in protecting against measles virus infection to an extent like that achieved by natural infection with measles virus, the vaccine did not cause any immunosuppression and left the children's defenses perfectly prepared to fight off other infections. These data indicate that vaccination against the measles virus not only protects against measles, but also helps maintain the effectiveness of other vaccines and builds up better defenses against the many infectious disease threats that threaten children's lives.

The above are not the only beneficial effects of vaccines, in addition to the protection against the disease they are intended to protect against. Another important indirect effect of vaccination is that it helps to slow the progression of antibiotic-resistant bacteria. The reason for this is obvious, since vaccines, by decreasing the incidence of infections, decrease the need to treat them with antibiotics, thus decreasing the likelihood of resistant bacteria. Be that as it may, research into the benefits of vaccines, in addition to probably putting them in their rightful place in the public eye, promises to reveal interesting secrets about how the immune system works and how it relates to cancer and other diseases.

## 7.- Evasion or death

As we have said, we currently have only 26 effective vaccines. One might ask why we have not been able to generate more, given the importance of vaccines in reducing child mortality in developing countries, as well as in making progress in eradicating serious diseases throughout the world. Could it be that not enough effort has been put into research?

In my opinion, and that of other scientists and reasonable people, investment in research could always be increased, and surely the profits made from that increased investment would more than compensate in the long run for the effort involved. For example, it would have taken much less money to prevent the crisis caused by the coronavirus SARS-CoV-2, by investing in the development of vaccines against this type of virus since the outbreak of the SARS virus epidemic in 2002, also caused by a coronavirus, than the money that will be needed to tackle it once it was triggered. Not to mention the losses caused by this pandemic. Without a doubt, if there is one research topic that can bring great benefits to all of humanity, it is vaccine research.

However, lack of research is not solely responsible for the fact that we lack vaccines for many diseases. The main culprits are, in fact, the microorganisms themselves, many of which have developed truly fascinating immune system evasion strategies that we will visit briefly in this section. This will give us a better understanding of why getting an effective vaccine for the majority of the population is not so easy, why it is probably not going to be easy to get for COVID-19, the illness caused by SARS-CoV-2 virus, and why an effective vaccine against the human immunodeficiency virus, HIV, which causes AIDS disease, has not yet been achieved and may never be.

### 7.1.- Mutation and evolution of the HIV virus

In fact, we are going to start with this last issue, because it introduces us in an excellent way to one of the fundamental issues that explain the resistance of microorganisms to the immune system: their evolution. Addressing what is known about this topic will also allow us to explore more deeply what the COVID-19 epidemic entails, which we will address in the next section, and discuss whether or not we will be able to develop an effective vaccine, and whether or not we will be able to have effective antiviral drugs.

To begin with, let's make it clear to anyone who might need it that evolution not only occurs with animal or plant species on a time scale of tens of thousands of years, but it also occurs with many microorganisms that infect us, and in the case of the HIV virus it occurs on a scale of only days or weeks. The HIV virus is one of the most rapidly evolving and adapting to changes in its environment, caused by the immune response or by treatment with antiviral drugs, within an infected person. SARS-CoV-2 probably also can evolve rapidly, as they can do all viruses whose genome is made up of RNA, although, at this point, we do not yet know how fast this virus may evolve.

The evolution of microorganisms, whose mechanism we will explain later, can help to clarify two rather surprising facts. The first, as we have already mentioned, is why it is difficult to generate an effective vaccine against the HIV virus and other RNA viruses, which also include the flu virus. The second is why the HIV virus generates resistance to antiviral drugs, but this resistance is not transmitted from person to person. In other words, a person infected with the HIV virus who receives antiviral drugs develops viruses that are resistant to them, which continues to infect them, although in a less virulent way than the initial one. If this person were to unfortunately infect another person, that person would not have drug-resistant virus at the time of diagnosis and could be treated with the same drugs that the person who infected him or her is treated with. This is not the case with antibiotic-resistant bacteria, which, when they become resistant to one of the antibiotics, can pass the acquire resistance on from generation to generation. Well, the evolution of HIV can explain both phenomena.

The above facts will take us into what I consider to be quite deep aspects of the mechanisms of natural evolution, aspects that will also allow us to understand how microorganisms generate and transmit resistance to the action of the immune system. To understand these profound aspects, it is convenient to start from the surface and briefly visit the biology of the HIV virus and how it generates the AIDS disease, the acquired immunodeficiency syndrome.

The viruses that cause AIDS, of which there are two main classes related to each other, HIV-1 and HIV-2, infect cells of the immune system that express on their membrane the CD4 co-receptor and the CCR5 chemokine receptor, to which they need to bind to be able to enter the cell and release their genetic material to reproduce. HIV-2 is mainly found in West Africa and is currently spreading in India. In the rest of the world almost all AIDS is caused by HIV-1, which is more virulent. Both viruses appear to have spread to humans from other primate species. According to the results of their RNA sequencing studies, it has been determined that HIV-1 has passed to humans on at least three separate occasions from the chimpanzee, while HIV-2 passed to humans from the mangabey, another primate.

Among the cells of the immune system that express CD4, in addition to CD4 T cells, of course, are macrophages and brain microglia cells, which are in fact innate immune system cells derived from bone marrow precursors. The infection of the virus eventually leads to the death of the infected cell. In particular, CD4 T lymphocytes die primarily from a type of cell suicide called **pyroptosis**. This is a process of apoptosis induced when a cell detects that it has been infected by an intracellular microorganism. To avoid spreading the infection, the cell secretes inflammatory cytokines as an alarm signal, after which it commits suicide. The cytokines help recruit immune system cells to the site of infection to try to fight it. In the case of HIV, however, the death of infected CD4 T cells eventually undermines the functioning of the immune system.

HIV infection from one person to another should be done through intimate fluid contact, which occurs only through sex, blood transfusion or the use of unsterilized hypodermic syringes used by several people for drug abuse. Saliva, sweat, or tears cannot transmit HIV.

The initial infection causes, two to four weeks later, a series of symptoms that can be mistaken for the flu. The immune system reacts within a few days against the virus and initially controls it, but is not able to eradicate it, as is the case with the flu virus. However, the symptoms disappear, and it seems that the person has recovered from the illness, but this is not the case. The virus continues to reproduce and infect new cells, particularly new CD4 T cells, but for reasons not yet well understood, this does not cause noticeable symptoms for several years. This period is called **clinical latency** of the HIV virus, and it can last from three to twenty years.

The clinical latency period is characterized by the fact that, although the virus continues to reproduce and its presence can be detected in the blood, the level of CD4 T lymphocytes remains adequate to keep the immune system functioning and to prevent other infections from developing. However, when the CD4 T lymphocyte numbers drops below 200 cells per microliter of blood, so-called **opportunistic infections** begin to occur.

Opportunistic infections are so called because they are caused by microorganisms or parasites that, in a person with a normally functioning immune system, cannot cause infection. However, as the number of CD4 T cells decreases, they are no longer able to provide the other effector cells with the stimulating and coordinating molecular signals that are necessary to properly control the infections. Among other effects, the decrease in $T_{FH}$ cells will affect antibody production, the decrease in $T_H1$ cells will affect macrophage activity, and the decrease in $T_H17$ cells will affect neutrophil activity. The entire adaptive immune system, which depends on the activation and differentiation of CD4 T cells, will be seriously affected, making the infected person susceptible to multiple life-threatening infectious diseases.

If opportunistic infections do not cause death, certain tumors that can develop as a result of the loss of a functional adaptive immune system may also cause death. These tumors include lymphomas and Kaposi's sarcoma, a tumor of the venous endothelial cells. In the case of AIDS patients, most of these tumors are usually caused by mutations following infection with other viruses, which can transform infected cells into tumor cells. If this happens, a normal immune system usually detects the tumor cells without major problems and eliminates them before they can generate a significant tumor. This elimination is usually carried out by cytotoxic CD8 T cells which, as we know, also require the collaboration of CD4 T cells for their complete activation. For this reason, if there are not enough CD4 T cells available, the transformed cells are not eliminated efficiently and are more likely to reproduce and generate a tumor that can be fatal.

As we can see, the HIV virus is not the direct cause of the death of the infected person, but it facilitates the entry and invasion of other infectious organisms, and it also facilitates the development of tumors. This effect, however, is not necessarily the most convenient for the virus to continue reproducing. Ideally, for any microorganism that infects an animal or plant, it should keep its host alive as long as possible, avoiding at the same time being eradicated. This would allow the microorganism to live a long time inside the host and thus have multiple occasions to infect others and spread through the host population, which ideally it would not kill, but only use for reproductive purposes. The HIV virus is very close to achieving the ideal state for an infectious microorganism, as AIDS disease takes several years to manifest itself and, in the meantime, the virus continues to reproduce in the infected person who is thus able to infect others for many years, until the immune system is finally irreversibly damaged by the virus. For some people, the virus can come even closer to the ideal state, or even reach it completely. It has been shown that about 5% of people infected with the HIV virus maintain high CD4 T cell levels without the need for antiviral drugs and take much longer than normal to develop AIDS, although they eventually do. However, 1 in 300 infected people maintain low levels of HIV in their blood and never develop AIDS. These people have been called "elite controllers" because they keep the HIV virus very well under control. These people have been found to possess MHC gene variants that are particularly effective in presenting HIV-derived peptides to T cells, resulting in a more effective than normal cytotoxic response capable of keeping infection at bay.

However, once infected, people cannot eradicate the virus despite having a normal immune system at the beginning of the infection. Why does this happen? As far as we know, the most important mechanisms to prevent the spread of a virus infection are the neutralization of the virus by the generation of antibodies and the death of the infected cells by pyroptosis or by the action of CD8 T cells and Natural Killer cells. Both mechanisms are necessary, and if one of them does not work effectively the other is unable to eradicate the infection by itself.

And indeed, the most effective of these two mechanisms, the generation of neutralizing antibodies, cannot stop HIV infection. A few weeks after the initial infection, patients undergo the so-called **seroconversion**, i.e. their serum is converted from the state of having no antibodies against the virus to the state of having antibodies that bind to the HIV particles. This seroconversion has traditionally been used as a diagnostic criterion to determine whether someone is infected with HIV, since the presence of antibodies does not, as in the case of other viruses, lead to the elimination of the infection. HIV can continue to infect new cells by binding to the CD4 and CCR5 receptors despite the presence of antibodies against it in serum and intercellular fluids. These antibodies are obviously not capable of effectively neutralizing it and do not prevent HIV from continuing to bind to cells expressing these receptors. The question is, why? How does HIV escape the neutralizing action of the antibodies?

HIV uses several molecular evolutionary strategies that help explaining this distressing situation. The first is that HIV mutates very frequently each time it reproduces, thus generating a population of viral particles with different epitopes. An antibody generated against one of these epitopes will not be able to neutralize all the viruses, since part of the population will have somewhat different epitopes. Moreover, the antibodies generated against these mutants may not have an adequate affinity to neutralize them.

Another strategy that HIV uses to escape the immune system is epitope suppression. HIV uses several strategies to achieve this. One is to place these epitopes in places that are inaccessible to antibody molecules, which would be too large to access. This is called **epitope occlusion**. Another strategy is to coat the potentially more vulnerable epitopes with carbohydrate molecules, the chemical nature of which generally does not allow the generation of antibodies with very high affinities for them. Remember that this strategy is also used by many bacteria, for which it is necessary to use conjugated vaccines if we wish to generate effective antibodies against them. Attempts have also been made to generate conjugate vaccines against HIV, but these have not been successful because HIV still has more strategies for evading the action of antibodies.

Probably the most effective strategy for evading the action of antibodies is the way in which HIV has evolved to decrease the avidity with which antibodies can bind to it. This strategy allows HIV to decrease the binding strength of IgG antibodies, which are the main class of neutralizing antibodies, by hundreds or thousands of times. Again, it should be remembered that avidity is high when several antigen-binding sites cooperate, thus keeping the antibody firmly attached to the antigen. Avidity depends on the number of binding sites that the antibody can establish with the antigen and is higher the more numerous these sites are. It should also be remembered that antibodies of the IgG class have only two binding points, separated by a variable distance that depends on the mobility of the two arms of the Y-shaped molecule, made possible by the hinge regions **(section 2.9.1)**.

In order for both binding points of IgG molecules to be able to bind to their epitopes, the antigen must have at least two repeated epitopes and these must also be located less than the maximum distance at which the arms of the antibody molecules can be separated, which is about 15 nanometers. If the epitopes are located further apart, the antibodies, despite having two attachment points, can only use one. If this happens, the continued binding of an antibody to the antigen will depend only on the affinity of the antibody, but it will not be able to dispose of the avidity provided by the binding of the two arms at the same time.

The affinity of the antibodies generated by the host cannot be controlled by any virus. It depends on the mechanisms of gene recombination and somatic hypermutation that the host sets in motion and that, currently, no virus is able to affect. However, the virus can control the avidity with which the antibody binds to it if it is able to separate its epitopes a greater distance than the two binding points that the antibody has in each of its arms. This is what HIV has achieved throughout its evolution. By decreasing the avidity of the binding, the antibodies cannot remain attached long enough to the HIV particles and these are thus not permanently neutralized. The phagocytes cannot capture and eliminate the virus particles effectively enough and they are more likely to infect other cells.

We should now take a brief look at the structure of the HIV virus to better understand how it manages to separate its epitopes so that they are further apart than the two arms of the antibodies. We can consider the HIV particle as a tiny sphere with a diameter of about 120 nm. This means that in just 1 mm there is room for about 8,300 viral particles in a single line. The structure of the sphere is maintained by a series of proteins generated from the virus genome, linked together as pieces of a spherical framework on which a lipid bilayer with viral and human proteins embedded in it is superimposed. The latter are proteins that the virus particles steal from the host cell when, after reproducing inside it, they kill it and leave it in search of other cells to infect.

Inside this tiny sphere are a number of viral proteins that are essential for the reproduction of the virus. Inside it is also the so-called capsid of the virus itself, which is composed of the union of many identical proteins that form a truncated cone shape. Enclosed in it there are two copies of the viral genome and the reverse transcriptase enzyme. As we have said, the HIV genome is not made up of DNA, but of single-stranded RNA. This RNA first needs to be copied into a double-stranded DNA, which must be integrated into the genome of the host cell before the virus can begin to reproduce. For this reason, the HIV virus belongs to the **retrovirus** family, which is made up of viruses whose genome is RNA, and which in order to reproduce need to first generate DNA from it, transmitting the genetic information "backwards" (*retro*). Once integrated somewhere in the genome of the infected cell, this DNA will behave as if it were a set of genes in the cell. It will generate RNAs that will function as messenger RNAs, which will be translated into the proteins that will form the new viral particles. Two copies of the RNA produced from the DNA integrated into the genome of the host cell will function as a genome and will be encapsulated in the capsid of the new generation of viruses produced.

Let us now return to the problem at hand, namely how the virus has been able to reduce the avidity that the host's neutralizing antibodies have against it. These antibodies, to neutralize the virus, must attach to the protein the virus needs to bind to CD4 and CCR5 molecules and thus infect the host cells. This protein was identified shortly after HIV was determined to be the cause of AIDS and has been called **gp120**. This strange name means nothing more than that this protein is a 120 kilodaltons (kDa) molecular mass **g**lyco**p**rotein.

The way in which pg120 is anchored to the surface of the virus is very important. This protein is non-covalently bound to the protein called **gp41**, which passes through the double lipid layer that the virus has stolen from the host cell and interacts with the proteins that form the spherical lattice, thus binding to it. This implies that the gp120 protein is not floating on the outer membrane and therefore cannot move freely from where it is.

This immobility of the gp120 protein is fundamental to diminish the avidity of the antibodies. If gp120 could move freely on the membrane, the distance between any two gp120 molecules would be variable and this would allow the antibodies to bind to two of them, one with each of their arms, when these molecules, in their free movement on the membrane, would be at a distance inferior to those 15 nanometers. This would allow to increase the strength of their binding to the viral particle. As the gp120 proteins are immobilized, this possibility does not exist.

However, in addition to not allowing mobility, another condition is still necessary to prevent antibodies from using their avidity to bind to two epitopes at the same time. This condition is that the distance between two gp120 molecules located on the surface of the virus must be greater than 15 nanometers. This is what the virus achieves by limiting the amount of gp120 proteins it places on the surface of its small sphere to a minimum. Each viral particle contains only about 15 copies of this protein, which are thus located at an average distance of more than 15 nanometers from each other. This minimum number of gp120 molecules could be even lower, but in that case, the virus would greatly reduce the probability of binding to a host cell to infect it. Therefore, the viral particles use the maximum number of copies of the gp120 protein that allows them to have a sufficient binding capacity to the cells to infect them and, at the same time, to be able to distribute them over an average distance greater than 15 nanometers, which is necessary to prevent the avidity effect of neutralizing antibodies.

Remember **(section 2.9.1)** that antibody molecules are subjected to continuous shocks from water molecules and other proteins found in blood plasma or intercellular body fluids. This molecular agitation gives them energy, which may be able to separate them from the epitope once they have bound to it. For this reason, that the antibodies have sufficient affinity, i.e. sufficient strength of attachment to the epitope on each of their arms, and that they have sufficient avidity, i.e. two or more attachment points to repeated epitopes, are both very important conditions in keeping the antibodies attached to their antigens. When the antibody does not have enough affinity, it needs to have sufficient avidity, and this is why IgMs are the first antibodies produced. Only when the affinity could be increased, in the process of somatic hypermutation, is it appropriate to produce IgGs or IgAs, which have fewer binding points, i.e. less avidity, but more binding force at each of those points, i.e. higher affinity.

We are now discovering that nature has developed antibodies with two arms because a single site of attachment is generally not enough to neutralize microorganisms by remaining attached to them. Fortunately, most microorganisms also need several repeated proteins located close together to generate enough avidity to bind to the receptors they need to infect the cells. For this reason, they cannot escape the neutralizing action of antibodies. However, the HIV virus has been able to generate proteins that bind to the CD4 and CCR5 receptors, which, with few sites of interaction, that is, without much need for the phenomenon of avidity, allow the binding to the cell and the entry into it. For this reason, the virus can afford to separate the gp120 binding proteins on its surface a greater distance than that required for the IgG antibodies to bind to them with their two arms and thus avoid being neutralized.

We also found that the structure of the antibodies is adequate to neutralize the microorganisms because, to remain infectious, the vast majority of them cannot place their repeated epitopes more than 15 nanometers apart. The epitopes are repeated because most, if not all, of the structural molecules, i.e. those that provide the basis on which the cells are built, are polymers of simpler molecules that are obviously repeated. In other words, living organisms cannot escape the conditions that make life possible, and these conditions involve the use of molecules formed by polymers that will necessarily show repeated epitopes at short distances.

However, another reason why microorganisms, particularly viruses, also need to have repeating epitopes is because these fulfil the mission of allowing interaction with molecules on the cell membrane that allow infection. For this reason, the most effective neutralizing antibodies are those that bind to these repeating epitopes. In addition, just as avidity is important in keeping antibodies bound to antigen, it is also important in allowing viruses to bind to various receptors on the cell membrane at the same time, or otherwise they could not stay bound to the cell membrane long enough to infect the cells. Since the protein molecules involved in these interactions must be within a certain size range, these repeating epitopes are located at distances typically less than 15 nanometers. For this reason, during evolution, antibody molecules have been selected so that the two arms that give them the ability to bind to these epitopes are separated by a maximum of that distance. In other words, throughout evolution, a maximum distance of 15 nanometers between the two antibody arms has conferred the greatest capacity for protection against practically all microorganisms and parasites that threaten us, due to their own vital needs for binding to the cells and infecting them. This is what has made it sensible for the immune system not to generate larger antibodies, which would be more costly. We thus see that the limitations of the microorganisms to infect the cells are used by the immune system to neutralize them. However, the HIV virus has been able to find an evolutionary gap through which it has seeped and has avoided the neutralizing action of antibodies.

The above considerations may be important in explaining why generating an effective HIV vaccine has not yet been achieved, and may never be, unless the human genome is modified to allow it to generate antibodies with longer, more flexible arms that can bind to epitopes more than 15 nanometers apart. As much as the vaccines that have been attempted to generate are capable of adequately stimulating the immune system and producing antibodies against the HIV epitopes, the problem is that the immune system cannot generate antibodies effective enough to neutralize the virus, because they cannot have sufficient affinity to bind to a single attachment point, nor are they large enough to bind to two attachment points at the same time.

This teaches us another important lesson about vaccines: they will only be able to generate protection if the immune system is sufficiently evolved to defeat the infection. If this condition is not met, it will not be easy to generate an effective vaccine, since the immune system mechanisms will be ineffective in defending us.

### 7.1.1.- Lost resistance

Now we must explain another very interesting phenomenon about the HIV virus, although it does not have a direct relationship with the immune system. This phenomenon is that the HIV virus becomes resistant to the action of the antiviral drugs and why this resistance is not transmitted to another person, even if this person is infected with a virus that is resistant to the drugs.

This situation contrasts with that of bacterial resistance to antibiotics, a growing global health problem. In this case, resistance acquired by a bacterium can be transmitted to the next generations of bacteria descending from the first one that acquired resistance. This is because resistance to a given antibiotic depends on a mutation in a bacterial gene, or on the acquisition of new resistance genes captured from other bacteria. These mutations and genes can be passed on to the next generations of bacteria, making them resistant.

However, in the case of the HIV virus, resistance to antiviral drugs resides exclusively in the generation of mutations in the genes of the virus. In this case, the generation of mutants is so dynamic and these change at such a speed that the mutations that allow resistance can be lost just as easily as they were acquired, if the environment in which the virus reproduces does not make it necessary for the mutations to be maintained.

In other words, due to the mechanism of HIV reproduction, especially the need to generate DNA from the two copies of RNA contained in the capsid, mutant viruses are generated in each of the infected cells. Thus, each patient does not have a virus, but a population of mutant viruses in his or her blood. These mutant viruses will compete to infect new cells. A reproductive race is thus established and those viruses that are most effective in reproducing will dominate the HIV virus population in an infected person.

For reproduction and infectivity, HIV viruses need two enzymes encoded in their genes: **reverse transcriptase and protease**. The reverse transcriptase is the enzyme in charge of generating the DNA copy from the viral genome RNA. The protease is in charge of generating gp120 and gp41 proteins from the precursor protein of both, called **gp160**, which is the protein produced from the messenger RNA generated after the integration of the viral DNA in the host genome.

Drugs that prevent the action of reverse transcriptase prevent the generation of the DNA necessary for the reproduction of the virus. Drugs that prevent the activity of protease prevent the generation of gp120 and gp41 proteins, which are necessary to allow the infection of cells by binding to CD4 and CCR5. Both drugs, therefore, prevent viral reproduction.

In fact, this is what happens initially in AIDS patients treated with them. The most competitive viruses in the absence of drugs are sensitive to them, i.e. they have very effective reverse transcriptase and protease variants, but are targeted by these drugs, which block their activity. However, the effect of the drugs is never total, as there are not enough of their molecules to find each reverse transcriptase and protease of all the viruses and block them. For this reason, some viruses continue to infect cells and continue to generate mutants. Some of these mutants will be less sensitive to drugs and will be the ones that manage to infect cells most effectively in the presence of these molecules. These infections may even generate new mutants until optimal mutants are achieved, i.e. those most resistant to the action of the drugs. These are the ones that are going to prevail in the virus population of an HIV-infected patient receiving antiretroviral treatment. These mutants, however, do not reproduce as rapidly as the original viruses, but they are the only ones capable of reproducing in the presence of the drugs.

Let us suppose that, unfortunately, a patient treated with antiretroviral drugs infects someone else. This person will suffer the initial symptoms of the infection, like those of the flu, as we have said. Already in this initial infection, the generation of viral particles is going to be very important, and hundreds of different mutants will be generated. Some of these new mutants will in fact have lost the mutations that made them resistant to the drugs, that is, they will have reverted to the original viruses. It also turns out that, in the absence of treatment with drugs, these original viruses are those that have the most effective reverse transcriptase and proteases for reproduction, and they are the ones that are going to reproduce most effectively in the new infected person. For this reason, they were the predominant ones in the initially infected person before he or she was treated with the drugs, and for this reason they will again be the predominant viruses in the newly infected person, unless he or she is treated with those drugs.

We see here the result of evolution by natural selection in real time. It is this evolutionary mechanism, made possible by the rapid generation and selection of mutants, that explains both the appearance of resistance to antiretroviral drugs and their disappearance as soon as the drugs are no longer administered, or the absence of resistance in people who have not yet received antiretroviral treatment.

### 7.2.- Multiple costumes

As is to be expected, the HIV virus is not the only microorganism that, in an involuntary but very effective way, has developed mechanisms in its evolution to try to circumvent the action of the immune system. A variety of bacterial species, protozoan parasites and other viruses now have interesting and clever mechanisms to evade the action of the immune system. Let's explore some of them.

As we know, microorganisms that reproduce in the extracellular regions of animal organisms, such as many bacteria, or that need access to them in order to infect the cells from there, such as viruses, are susceptible to the neutralizing and opsonizing action of antibodies. For this reason, many of these microorganisms have developed strategies that modify the epitopes capable of generating an antibody production response. One of these mechanisms by which microorganisms vary their epitopes is called **antigenic variation**.

An example of antigenic variation is shown by the bacterium *Streptococcus pneumoniae*. This bacterium normally inhabits the epithelial surfaces of the airways, nasal cavities and lungs, but it causes disease only in people who are immunosuppressed in some way, as well as in the elderly and children. It is the main cause of pneumonia in these people and can also cause meningitis. It is also the leading cause of septic shock in AIDS patients, who are susceptible to widespread infection.

More than 90 different types of *S. pneumoniae* bacteria have been identified according to their ability to generate antibodies in infected people. Because the types have been identified according to the antibodies present in the serum of these people, they have been called **serotypes**. Each bacterial serotype generates an antibody that can only opsonize that particular type, but cannot opsonize other types. Thus, an effective immune response against one type does not provide protection at all against other types of the same bacteria, which can then infect the same person.

This capacity of S. pneumoniae depends on its ability to generate external polysaccharide capsules, i.e. different carbohydrate units, which act as different "disguises" for the bacteria. In other words, the more than 90 different serotypes of this bacterium are shown to the immune system as if they were different bacterial species thanks to their different outer layers that encapsulate and protect them. Thus, the primary response of antibody generation is only able to neutralize one of the types and, more importantly, to generate immune memory against that type only. This leaves the body unprotected against other types of *S. pneumoniae*, which for this reason can cause disease multiple times in the same person.

This immune system avoidance mechanism depends on the collaboration of the different *S. pneumoniae* serotypes to transform into each other through DNA transmission between them. This process is called **bacterial transformation**, and *S. pneumoniae* bacteria use it very effectively to evade the action of the immune system, as well as to transmit genes for resistance to different deleterious factors, including some antibiotics.

Another microorganism that uses the mechanism of antigenic variation in a more dynamic and dangerous way than the previous one is the flu virus. Influenza is a disease that appears in annual outbreaks, usually in winter in the northern or southern hemispheres, although near the equator outbreaks can occur at any time of year. The virus causes several million severe cases worldwide, resulting in up to half a million deaths each year, depending on the variety of viruses generated and their virulence. This is not surprising, because the virus infects 20% of non-vaccinated children and 10% of non-vaccinated adults worldwide each year. This means that having overcome flu illness one year does not ensure that your immune system is protected against virus outbreaks in later years. Why does this happen? To understand this, we must briefly look at the mechanism of flu virus replication.

First, as with all viruses, the influenza virus must bind to a molecule on the host cell membrane and use it as a gateway to the cytoplasm and cell nucleus. For this task, the virus has a protein called **hemagglutinin** (because in addition to allowing the virus to infect the epithelial cells of the lung and airways, it is also able to bind to the surface of red blood cells). This protein binds to carbohydrates linked to some glycoproteins on the membrane of lung, nose or throat epithelial cells. The binding automatically triggers a mechanism (called endocytosis) that forces the cell to incorporate the viral particle into its interior, where hundreds of new viral particles are generated using the cell resources.

When it detects the infection, the immune system sets in motion the proper mechanisms to defend against it. One of the most effective and absolutely necessary to prevent the progression of the infection is the generation of virus-neutralizing antibodies. These bind to the viral haemagglutinin and prevent it from binding to the surfaces of the cells that the virus infects. Thus, unable to enter the cells, the virus cannot replicate and is furthermore phagocytized by the cells of the immune system that detect the antibodies attached to it. The infection is thus defeated, and memory B cells are also generated which can produce antibodies against the same virus more quickly in the event of a possible recurrent infection. However, as we said, this immunological memory, in this case, does not ensure effective protection against viruses that will appear in subsequent years. The reason for this is the high rate of mutation that this virus undergoes in the process of reproduction in the many infected people. This causes some mutant viruses to vary their haemagglutinin gene so that the antibodies generated againts previous viruses are no longer able to neutralize the new virus particles, even though the mutated haemagglutinin is still able to bind to the proteins of the cell membrane and initiate an infection.

Like the HIV virus, the flu virus also has an RNA genome, which is particularly susceptible to mutation. In fact, I should say the flu viruses, as there are four types of these, A, B, C and D. The first three types can infect humans, and although infection by the last type, which infects pigs and cows, has not been detected, it cannot be ruled out.

Despite being a virus that has a genome made up of RNA, the mechanism of replication of this virus is different from that used by HIV. The flu virus does not need to transform its genomic RNA into DNA. The genomic RNA is used directly by a viral enzyme to generate new copies of RNA. In other words, RNA doesn't need DNA as an intermediary to copy itself. Some of the copies will be used as messenger RNA to generate the proteins of the viral particle and others will be used as the genome for the new particles.

The viral enzyme (called RNA-dependent RNA polymerase) that generates the new copies of RNA is not a reliable enzyme and, in the copying process, makes mistakes and changes some "letters" for others. This generates mutant viruses that are slightly different from the original.

Several types of new viral particles can be found among these mutant viruses. Some of these may be non-viable particles, i.e. those in which the mutations that have occurred prevent them from infecting and reproducing. Other types of mutant particles include those that have mutated other genes, but not the haemagglutinin gene, or at least not enough so that the antibodies generated in a previous infection can no longer neutralize these new particles. Finally, it is also possible that some newly generated viral particles have mutated haemagglutinin so that it can continue to bind to the cells that the virus infects, but to which the antibodies generated in a previous infection, and still present in the blood and tissues, can no longer effectively neutralize. These new viral particles will be infectious and will have eluded the immune memory. This process of normal mutation of the influenza virus that occurs as a result of its reproduction is called **antigenic drift**.

Fortunately, the dynamics of annual influenza virus infections means that most people have been infected at least once in their lives and have generated an immune memory, i.e. they have antibodies circulating in their blood and body fluids against the variant of the influenza virus that infected them and can generate, thanks to the memory cells, more antibodies quickly against that particular virus. These antibodies are directed not only against the haemagglutinin of the virus, but also against other viral proteins. Thankfully, there is also a limit to the extent to which the virus can mutate. Mutations must not disable haemagglutinin in its ability to bind to cell membrane proteins, or the virus will no longer be able to reproduce. For this reason, most of the mutant viruses that occur each year are still partially neutralized by the antibodies present in the majority of the population, and the population can also respond with its memory cells, which will detect the epitopes of the mutated virus, despite the mutations. However, this neutralization and memory response occurs at different degrees of effectiveness, depending on the B cells and the specific antibodies that everyone possesses and how these fit more or less tightly into the mutated epitopes. However, each flu season a small percentage of the population will either never have been infected with influenza virus and may be infected by it, such as young children, or if they have overcome influenza caused by a particular mutant in previous years, they will not have antibodies effective enough to neutralize the new mutant virus that has arisen that season.

This explains why the flu vaccine has to be produced and given every year. Normally, an influenza vaccine given in a particular year does not ensure effective protection against mutant viruses that will occur the next influenza season, and the vaccine loses all effectiveness in just a few years. Vaccines should also be prepared well in advance with the mutant virus variant detected at the time when it becomes necessary to start this preparation. Detecting or predicting the influenza virus variant for the next season is a responsibility of the World Health Organization. This information is communicated to the pharmaceutical companies that prepare the vaccine, but this must be done at least six months before the start of the flu season, which begins in December or January in the northern hemisphere, so that these companies have time to prepare the millions of doses of vaccine that must be administered. During that time, the detected virus variant, which is expected to be the predominant next flu season, may mutate into a new variant for which the vaccine prepared that year may not be effective. This happens from time to time and, in fact, it happened in 2018.

The above mechanism of mutation also explains why two winter influenza epidemics occur each year, one in the northern hemisphere and one in the southern hemisphere. Fortunately, epidemics usually affect only a limited percentage of the population, as we have explained. Nevertheless, millions of cases still occur every year, which is enough to qualify the situation as an epidemic.

Sometimes, however, so-called influenza pandemics can be triggered, in which the variant of the virus produced can infect a higher percentage of the population and thus spread more quickly. These pandemics are not normally generated by the process of antigenic drift mentioned above, but are produced by a more perverse mechanism in which two very different viruses, perhaps one that infects humans and the other that infects pigs, mix their genomes at random and generate, by bad luck for us, a new infectious virus for which the majority of the population lacks antibodies and memory cells that can effectively neutralize it and react against it. This mechanism is called **antigenic shift** and is more drastic than the previous one, since the new virus is not only a mutant, it is a hybrid between two viruses. This hybrid virus can occur if an animal or a human being is simultaneously infected by two viruses of two different variants. Simultaneous infection makes it possible for some cells to be infected by both viruses at the same time, and when they reproduce simultaneously, they mix their genomes and generate completely new viruses. The probability of this happening is not very high, but it increases in environments where there is close contact between humans and farm animals that may be infected by influenza virus. This makes it possible for the new virus to infect humans at the same time as the seasonal virus produced by antigenic drift.

Thus we see how the mechanisms inherent in the evolution of species, the generation of mutants and their selection in a given environment, in this case the human population and its neutralizing antibodies, is taken advantage of by the influenza virus to find itself in a state of continuous evasion of the human immune system. This system, in the global population, is always chasing the virus through the space of mutations that it occupies, always trying to neutralize it, in general achieving it, but its objective being frustrated later by the continuous generation of mutant virus occurring in a greater or lesser extent in each infected individual.

More evolved and complex organisms than viruses can employ the mechanism of antigenic shift in a more sophisticated way. This is the case with the protozoan *Trypanosoma brucei,* which causes sleeping sickness. Throughout its evolution, this microorganism has generated a genetic mechanism specifically designed to evade the action of antibodies. Let's see how it works.

*Trypanosoma brucei* has its surface covered by about five million molecules of a protein called variant surface glycoprotein, or VSG. This protein is strongly immunogenic, that is, it induces a strong immune response against it, in particular, it induces the generation of antibodies, which will neutralize, opsonize and activate the complement and will quickly eliminate the trypanosomes that express that VSG variant. All of them? No! A small percentage are capable of changing the VSG protein for another variant to which the antibodies generated cannot bind. This small percentage of trypanosomes are now immune to the action of the antibodies and reproduce rapidly. The immune system, however, detects, as it has done before, the new variant of the VSG protein and generates antibodies against it again. These new antibodies are also effective in neutralizing and opsonizing the trypanosome, but again a percentage of these escape the action of the new antibody by changing the VSG protein to yet another different variant to which the previous antibodies cannot bind. The trypanosome reproduces again without impediment until a new antibody is generated against this third VSG variant. These cycles continue until death occurs. Continued cycles of inflammation in response to cycles of antigenic variation eventually compromise blood circulation and can cause irreversible failure in a vital organ, particularly the kidney or the brain.

As we see, paradoxically, this mechanism involves inducing the immune system to produce an intense antibody response against variants of a particular antigen of this protozoan, and to change this antigen to another when the antibodies have been produced. This genetically directed antigenic variation renders the former antibodies useless and forces the immune system to produce more antibody against the new VSG. The mechanism of variation occurs because the genome of this microorganism has about two thousand VSG variants. A complex mechanism ensures that every so often, similar to the time required for the immune system to generate antibodies, the DNA fragment corresponding to one of these variants, chosen at random, is copied and transferred to the site of the trypanosome genome that directs the production of VSG messenger RNA. In this way, the site of messenger RNA production is occupied by different variants of the VSG gene on a regular basis. This periodicity keeps the immune system in a state of "constant pursuit" of the microorganism, without ever being able to reach it and eliminate it completely.

The cycles of antigenic change generated by the parasite help explain the cycles of drowsiness and wakefulness caused by trypanosome. It produces a substance, called tryptophol, that induces sleep. At the peak of the infection, when the immune system has not yet produced the first antibody, trypanosomes are abundant in the blood and tissues, so they produce high amounts of this substance. When the antibody is produced in sufficient quantity and begins to eliminate the trypanosomes, the amount of tryptophol decreases, and the infected person momentarily leaves the sleep state. Changing from one VSG variant to another allows the trypanosome to escape the action of the antibody and begin to reproduce again, so the amount of tryptophol produced increases and the infected person falls back into a state of drowsiness, and so on.

### 7.3.- More microbial tricks

In addition to the "masquerade switching" technique mentioned above, microorganisms have an arsenal of other mechanisms to slow down the action of the immune system against them. These mechanisms, let's not lose sight of it, have been developing along the joint evolution between the microorganisms and the organisms they infect. If the latter have been developing increasingly powerful and sophisticated immune systems and infection eradication mechanisms, the others have been "learning" how to bypass these mechanisms or even manipulate them to their own advantage. There is no trick that microorganisms do not try to use to evade the action of the immune system, or even to use it for their own benefit. No matter how spectacular and effective the weapons used against the enemy may be, the enemy can also use them against us. Let's explore some of the tricks used by microorganisms to evade the immune system, which can only be understood in light of how the immune system works.

Let's start with one of the most cunning microorganisms to escape the immune defenses: The bacterium *Staphylococcus aureus*. This bacterium uses several strategies against the cells of the immune system. One of them is that it can let itself be phagocytized by macrophages, but avoid being digested by them, so the bacteria lives inside these cells. When the bacterium detects a favorable change in conditions, it reproduces inside the macrophage, kills it, and starts another infection. *S. aureus* is equally capable of secreting substances toxic to neutrophils because the death of many of these cells benefits the progress of the bacteria.

In addition, *S. aureus* is able to use DNA from the molecular net secreted by neutrophils **(section 2.8.1)** to produce toxic substances. This bacterium produces two enzymes that degrade the DNA of the molecular net and transform it into a substance that induces macrophage apoptosis. These two enzymes catalyze chemical reactions in the DNA net. The first enzyme is **Staphylococcus nuclease**, which acts on the DNA and breaks it down into its basic molecular constituents, the individual "letters", thus breaking down much of the molecular network and rendering it useless. The second enzyme is called **adenosine synthase**, which acts on one of the components of DNA (adenine) and converts it into a substance called 2'-deoxyadenosine. This is the trigger for the suicide of macrophages.

Another bacterial strategy that fortunately does not always work is that used by the mycobacterium that causes leprosy, *Mycobacterium leprae*. This bacterium infects tissue macrophages and nerve cells and lives inside them until it kills them. Antibodies are not effective against them, as these molecules cannot be internalized by the cells. The only effective way to fight intracellular mycobacteria is their phagocytosis by macrophages or neutrophils and their destruction inside these by respiratory burst **(section 2.2)**. Remember that this requires the production of IFN-$\gamma$ and the expression of CD40L by $T_H1$ cells. In the absence of generation of this class of helper T cells, *M. leprae* infection will be much less well controlled. In some cases, this mycobacterium can confuse the immune system into believing that it is an extracellular microorganism. Thus, the immune system activates the mechanisms and weapons needed to fight against extracellular, not against intracellular organisms. In these situation, $T_H1$ cells are not generated. Instead, $T_H2$ cells (another type of CD4 T cells that help B cells to generate antibodies of the IgE class) or $T_{FH}$ cells are produced. In the lack of $T_H1$ cells, macrophages cannot be stimulated through CD40L and the generation of IFN-$\gamma$ and the respiratory burst in the infected macrophages cannot take place. In addition, $T_H2$ cells secrete cytokines that inhibit the generation of $T_H1$ cells, so once the immune system is induced to generate $T_H2$ cells it is virtually impossible for it to change its mind and generate $T_H1$ cells.

The unfortunate cases of *M. leprae* infection that lead to the incorrect induction of $T_H2$ cells result in a severe type of leprosy disease, which has been called **lepromatous leprosy**. In this form of leprosy, patches of skin, papules and nodules, called lepromas, appear. The progression of the disease without proper control by the immune system results in severe tissue destruction, particularly in the cartilage of the nose and ears, which deforms the face. In advanced stages of the disease, the so-called lion face appears, characterized by a deformation of the face that vaguely resembles that of a lion because numerous lepromas have developed and spread over the entire face. There is also sensory loss due to damage to nerve cells. The disease is much less severe, but still serious, when the immune system of the infected person reacts correctly. In this case, the variety of the disease is called **tuberculoid leprosy**, where the areas of infection show many macrophages and neutrophils and far fewer bacteria, although the collateral damage caused by the respiratory burst in these tissues is high.

We now turn to another strategy that some viruses and some bacteria can use to avoid detection. This strategy is called **latency**. Latency is a state in which an infectious organism, after penetrating into a cell and infecting it, does not actively reproduce until conditions warrant it. These conditions depend to some extent on the health of the host cell and the general health of the organism, especially the general state of immunocompetence, i.e. how effective the immune system can be at any given time in fighting it. In some way, latent microorganisms detect when the immune system is at its worst, perhaps because it has had to fight another infection, or because it is not properly nourished, and at that moment it triggers the process of its reproduction. Under conditions of reduced immune competence, the microorganism will have a greater chance of not being eradicated before it can reproduce sufficiently to infect another organism.

Latency is an effective mechanism of resistance because, let's remember briefly, to detect the presence of an infectious microorganism the immune system needs to process it and generate peptides that must be presented on MHC molecules. The microorganisms acquire the latency state after infecting a cell and prevent the cell from destroying them and digesting them into peptides. Since they do not reproduce, they do not generate proteins either, so the cell cannot indicate that it is infected since it is unable to present peptides derived from proteins of the microorganism on the MHC molecules. Thus, the microorganism goes unnoticed by the immune system.

Many viruses acquire a latent state by the strategy of integrating their genome into a chromosome of the infected cell. Once integrated there, the virus genome remains inactive until some signal indicates that it is a good time to activate it and move to the reproductive state, generating viral proteins to form new virus particles. We have already seen that the HIV virus integrates its genome into the cell it infects and, indeed, in some of these cells, particularly dendritic and macrophage cells, it can acquire the dormant state. This is yet another reason why the immune system could not completely eradicate it, even if it could produce more effective antibodies against it.

Viruses that frequently infect us and use latency to evade the immune system include viruses in the herpes virus family. Among these is the common herpes virus that produces the lip lesions that can appear during or after colds and that are produced by this virus and not by the cold virus. The common herpes virus infects the epithelial cells of the lip and face, which it damages. The immune reaction against these infected cells is what generates the inflammation and the lip lesion, which disappears after a few days. However, the virus can infect not only the epithelial cells, but also some neurons. From the lip it passes to the nerve endings that innervate it and establishes itself in the neurons of the trigeminal ganglion, which is located at the level of the ear. In the neurons, the virus acquires a latent state, which is also favored by the fact that the nerve cells express low amounts of MHC-1 molecules, because they are cells that should not be easy targets for cytotoxic cells. Killing neurons, even if infected, always poses a greater risk than killing another class of cells, so before killing them it is necessary to ensure that it is the only way to go.

Another virus that accompanies practically all of us and that also uses latency, among other escape mechanisms, as a means of resistance against the immune system, is the **Epstein-Barr virus** (called **EBV** virus), which also belongs to the herpes virus family. I had the opportunity to meet Dr. Sir Michael Anthony Epstein personally while I was doing my doctoral thesis in France, on the occasion of the award ceremony of a prize granted for his contribution to the discovery of this virus and the role it plays in the development of the so-called Burkitt's lymphoma, which the virus is capable of causing in regions where malaria is endemic. It seems that the dual infection with malaria and EBV virus results in a significant increase in the incidence of this type of tumor.

It is estimated that about 95% of the human population is infected with the EBV virus, which causes the disease called **infectious mononucleosis**. Normally, the infection occurs in early childhood, after breastfeeding is stopped and children are no longer protected by the antibodies that the mother passes on to them with her milk. The virus infects B lymphocytes and epithelial cells and generates an immune response against it. However, when the infection occurs in childhood, the immune response is not very strong and does not generate serious symptoms. In countries where hygiene conditions are very high, as is generally the case in many regions of developed countries, infection may be delayed until adolescence, when the disease is usually contracted through the first kissing on the mouth. For this reason, this disease is popularly known as the **kissing disease**. In this case, the immune response is stronger and generates symptoms that can be severe, including fever, sore throat, significant swelling of the lymph nodes in the neck, and tiredness. In the most severe cases, liver and spleen swelling may also occur.

After the initial infection, the virus enters a latent state inside the cells it has infected. This state prevents the activated cytotoxic T lymphocytes, generated in response to the infection, from identifying the infected cells, so the virus remains hidden in the body, waiting for some opportunity to reproduce. This opportunity may come from a decrease in the fitness of the defenses, or from changes in the metabolic state of the cell, which the virus detects in some way, after which it reactivates. Under normal conditions, there seem to be some reproduction peaks that, nevertheless, are usually finally controlled by the activation of the cytotoxic memory T cells. These cells are activated and detect the infected cells, killing them. The organism is thus in a stalemate situation with the virus, unable to defeat it completely, but not letting it defeat us either.

However, the EBV virus is not content to just wait for opportunities that cause immunosuppression. The virus has a number of genes that, when activated, help some of the B cells it has infected to proliferate, increasing the likelihood that they will become tumor cells. The lymphocytes in a state of continuous cell reproduction are the ones that the virus needs most to reproduce more quickly. In general, viruses do not reproduce in quiescent cells, i.e. those in a state of suspended reproduction. Thus, the transformation of some B lymphocytes into tumor cells also stimulates the reproduction of the EVB virus in them. These tumor cells, however, will present virus peptides in their MHC-1 and can therefore be eliminated by the cytotoxic T cells. This is the reason why, even though most of the population is infected by the EBV virus, few people develop leukemias or lymphomas, which are the main types of tumors induced by this virus. However, if the immune system is unable to kill the transformed cells, lymphomas and several other tumors, such as gastrointestinal cancer and nasopharyngeal carcinoma, can arise.

In addition to latency, as we have already mentioned, microorganisms use other strategies to subvert the immune system and prevent it from functioning normally. In fact, there may be no mechanism of the immune system that is not manipulated in one way or another by one microorganism or another. In particular, viruses whose genome is made up of DNA, not RNA, are those that have developed the most strategies to subvert the immune system. The reason is that this type of virus cannot mutate as quickly as viruses whose genome is made up by RNA and cannot use this mechanism to evade the immune system. However, the lower mutation rate allows these viruses to have larger genomes than

RNA viruses. The reason for this is that a high mutation rate is more likely to render a large genome unusable by accumulation of errors than a small one. These larger genomes allow DNA viruses to have a much larger number of genes than RNA viruses and, among these genes, they have some that act on the mechanisms of action of the immune system. In fact, these genes are so important for the survival of DNA viruses that up to half of the genome of some species of this class of virus is dedicated to containing genes that subvert the immune system. These genes have been captured by viruses throughout evolution, from the very genome of the organisms they infect. Some of these genes function as oncogenes, that is, they stimulate the formation of tumors, which favors the reproduction of the virus, as we have already said. Other genes captured, however, are genes that produce proteins interfering with the mechanisms of the immune system that aim to control the infection of viruses, especially the activity of cytotoxic CD8 T cells and Natural Killer (NK) cells, which we will discuss in more detail later. Some viruses are also capable of tricking the immune system into believing that the infection has been stopped, although in reality it has not, and they set in motion the regular mechanisms to slow down the immune response once the infection has been beaten, which obviously leads to a continuing infection and to the inability to beat it.

One of these genes is the one that produces interleukin-10 (IL-10). This cytokine generates various anti-inflammatory effects, i.e. it slows down the immune response. Among them, there is a decrease in the secretion of cytokines by $T_H1$ cells, and a decrease in the expression of MHC-2 molecules and co-stimulatory molecules by macrophages. These effects decrease the activation of $T_H1$ cells and thus their ability to activate CD8 T cells, allowing the virus a better chance of survival.

Other genes act on the activation of the inflammatory response, which is always necessary to stimulate the defense mechanisms, or act on the defense mechanisms themselves and their coordination. Among the genes stolen from their hosts by various species of viruses are genes for receptors of various cytokines, chemokines, complement or even antibodies. These genes produce soluble proteins that will be secreted to the exterior by the infected cell. As they are receptor proteins for various other proteins, and are free in the external environment, they will bind to these and thus prevent them from binding to their true receptors, which are present on the external membrane of the cells of the immune system. For example, a viral receptor of a cytokine will associate with it and thus prevent this cytokine from binding to the true receptor expressed on the surface of an immune system cell, precluding it from being correctly activated or acting. Among the soluble receptors produced by some viruses are those for interleukin-1 (IL-1), TNF-$\alpha$ and IFN-$\gamma$. These cytokines are important stimulators of inflammation and the fight against viruses.

Other soluble receptors produced by viruses that block the action of some important components of the immune system are the Fc and complement receptors. By binding to antibodies, soluble Fc receptors block their effecting function, for example, by preventing them from activating complement, which will prevent opsonization of virus particles and their removal by phagocytosis, or by preventing phagocytosis mediated by Fc receptors expressed by phagocytes and which capture microorganisms opsonized by antibodies. Fake complement receptors will also associate with some complement components and block their effector function or activation, in particular by blocking the uptake of opsonized virus particles by cells expressing true complement receptors.

### 7.3.1.- Micro RNAs

DNA viruses are also capable of generating so-called microRNAs (miRNAs). These are small fragments of RNA, about 19 to 22 "letters" (nucleotides), whose sequence is complementary to the "letter" sequence of one or another messenger RNA (mRNA) produced by a gene. Normally, these miRNAs bind to the tail of a particular mRNA (to its 3' region), which destabilizes it and prevents it from being translated into protein.

Again, the viruses that most use this strategy to prevent translation to proteins from mRNA produced by genes involved in the antiviral response are viruses of the herpes family; in particular, the Epstein-Bar virus. EBV encodes at least 44 miRNAs that are theoretically capable of preventing or at least slowing down the functioning of hundreds of genes. These miRNAs are produced by the virus at virtually all stages of its complicated reproductive cycle and can affect both innate and adaptive immune system cells.

The knowledge accumulated to date indicates that one of the functions of the miRNAs produced by the EBV virus is to promote the survival of EBV-infected B cells. This makes sense, since if the EBV-infected B cell dies, the virus dies with it. For this reason, these miRNAs also promote the growth of B cells that have become malignant, aided by the activity of the virus.

However, many of the miRNAs generated by the EBV virus are also intended to evade the action of the immune system. These miRNAs can attack the action of $\alpha$ and $\beta$ interferons, which are important molecules produced and secreted in response to virus infection, and which trigger a variety of effects to make it difficult for them to reproduce. The miRNAs can also affect the activity of certain cytokines important for the adaptive cellular response and the activation of $T_H1$ and CD8 T cells, or the chemokines necessary for the mobilization and organization of defense mechanisms. Similarly, miRNAs can affect the activity of IFN-$\gamma$ and even the activity of Toll-like receptors specialized in the detection of viral molecules. Finally, they can also affect the activity of NK cells, which we will discuss in detail later.

### 7.3.2.- Immunoevasins

However, the most important cellular mechanism on which viruses act to escape from the immune system is the presentation of peptides by MHC-1 molecules. If this presentation is inhibited in one way or another in an infected cell, cytotoxic CD8 T cells will not be able to detect the infected cells and will not be able to induce their apoptosis by perforins and granzymes. For this reason, many DNA viruses have genes that produce one or several proteins called **immunoevasins**, which prevent the presentation of peptides on MHC-1 molecules. This, in fact, leads to the lack of MHC-1 expression on the surface of these infected cells, since, without peptide binding, MHC-1 molecules are unstable and are directed to degradation. Again, throughout the mutual evolution of the virus and the animal immune systems, viruses have been able to acquire inhibitors for each of the steps that the cells use to process protein peptides produced in the cytosol and bind them to MHC-1 molecules.

We have not yet discussed this mechanism, which corresponds to the second phase of the overall process of loading peptides onto MHC-1 molecules after the proteasome (**section 5.2.2.2**) has degraded the proteins in the cytosol and generated the peptides that must be bound by MHC-1 molecules for transport to the outer membrane of the cell. We will briefly explain this below, which will allow us to better understand the tricks used by viruses to block it.

The molecules that the cell needs to transport to the outer membrane, such as MHC-1 molecules, are synthesized in a system of vesicles formed by a lipid bilayer similar to that of the outer membrane, called **the endoplasmic reticulum**, which is one of the most important cell organelles. The protein molecules that must be placed on the membrane or secreted to the exterior are synthesized within this system. The endoplasmic reticulum membranes fuse with the outer membrane and thus place in that membrane or secrete the synthesized proteins inside. This is how the cell has solved the problem of separating the proteins that should be destined to the outside from the proteins that should remain inside the cell.

However, the solution to the above problem has created another one. This is that the peptides produced in the cytosol thanks to the action of the proteasome must be transported to the interior of the endoplasmic reticulum, where the MHC-1 molecules have been synthesized. These molecules must carry self or foreign peptides before being transported to the exterior membrane. This implies that the peptides produced in the cytosol must be helped to pass through the lipid bilayer of the endoplasmic reticulum, since peptides, being of a hydrophilic nature because their ends are always charged, cannot pass through it without being transported through it. This transport is fundamental, because MHC-1 molecules that for some reason cannot properly load peptides are unstable and are transported to the interior of the cell, not to the exterior of it, for their degradation.

The transport of peptides from the cytosol to the endoplasmic reticulum must be done in an active way, that is, through the action of proteins that capture them in the cytosol and introduce them into the endoplasmic reticulum thanks to the contribution of metabolic energy. There are two transporter proteins that act together, called **TAP-1** and

**TAP-2** (Transporter-associated with Antigen Processing 1 and 2). These proteins are inserted into the membrane of the endoplasmic reticulum and pass through it together, forming a channel through which they actively introduce peptides from the cytosol. The part of these proteins found on the cytosolic side of the membrane has an area called the ATP-binding cassette. This zone is capable of binding to the molecule that all cells use as a source of metabolic energy, adenosine triphosphate (ATP). The hydrolysis of ATP provides the energy necessary for the peptides captured in the cytosol by these cassettes to be translocated to the interior of the endoplasmic reticulum, where the MHC-1 molecules that are being actively synthesized there are found. This allows the peptides generated in the cytosol by the proteasome to be loaded into the peptide-binding cleft of MHC-1 molecules.

Now that we know the basis for the transport of peptides produced by the proteasome from the cytosol to the endoplasmic reticulum, we can better understand how viral immunoevasins work. One of these molecules is capable of interacting with the ATP-binding cassette, which is the entrance door to the peptides, and blocking it, preventing them from penetrating into the endoplasmic reticulum. We have already said that, in the absence of peptides, MHC molecules are unstable and are degraded, so the cell infected by virus with this type of immunoevasins will show a lower than normal number of MHC-1 molecules in its outer membrane. This will make it very difficult for CD8 T cells that may have been generated in response to the viral infection to detect and eliminate the infected cells, which will prevent the infection from being eradicated.

The peptides that are transported to the interior of the endoplasmic reticulum are not charged to the MHC-1 molecules by mere diffusion. On the contrary, there is a complex formed by three different proteins, called the **peptide-loading complex**, which participates in the binding of the peptides to the MHC-1 binding cleft. There is even an enzyme that cuts peptides too long, which cannot be bound by MHC-1 molecules, to reduce their length so that they can fit into the binding cleft.

The peptide-loading complex functions as a safety mechanism. This protein complex keeps MHC-1 molecules in a semi-stable state and retains them in the endoplasmic reticulum to allow a reasonable time for one peptide or another to bind strongly to the binding cleft. The complex only releases the MHC-1 molecule when it has captured one peptide strongly and does not release it. If after binding the peptide is released this is because the bonding with the MHC-1 molecule was not stable and it must wait until another peptide of those transported to the endoplasmic reticulum is strongly and stably bound. The binding of the peptide must be very strong, because once outside the cell the peptide must not be separated from the MHC-1 molecule, and this for two reasons. The first one is that, if the peptide were to separate, the MHC-1 molecule would become unstable and would be degraded. The second reason is that the identity of the cells, their health status, that is, whether they are completely trustworthy or have been subverted by a virus, depends on the fact that peptides derived from cellular proteins are continuously presented by MHC-1 molecules. Without peptides stably bound by these molecules on the surface of the cells they cannot indicate their identity or their health status correctly and will be eliminated by the Natural Killer cells, as we will see in the following section.

It is not surprising, therefore, that some viruses produce proteins that interfere with the loading mechanism of peptides to the binding cleft of MHC-1 molecules. Since as long as these peptides are not charged, the cell does not allow the transport of MHC-1 molecules to the surface, these infected cells will be prevented from presenting peptides, will also have few MHC-1 molecules on the surface, and will not be able to be properly eliminated by CD8 T lymphocytes.

Although many species of DNA viruses have evasion mechanisms that affect the presentation of peptides by MHC-1 molecules, viruses of the herpes family are those that throughout evolution have acquired the most extensive arsenal of tools to decrease this presentation. This family of viruses possesses genes that produce proteins that prevent the presentation of peptides by MHC-1. One of these proteins promotes the degradation of MHC-1 molecules. This mechanism of degradation is called **dislocation**, because it dislocates, that is, it places MHC-1 molecules out of their normal place. The viral molecules that achieve this are ubiquitin ligases that viruses have stolen throughout the evolution of the hosts they infect. These ubiquitin ligases will bind ubiquitin to the MHC-1 molecules, which will direct them to the proteasome for degradation. Paradoxically, this mechanism destroys in the proteasome the proteins that require of this organelle to perform their function.

Other molecules produced by some viruses manage to retain the MHC-1 proteins at the membranes of the Golgi apparatus and prevent them from being exported to the external membrane. The Golgi apparatus is the organelle in charge of collecting the molecules synthesized in the endoplasmic reticulum and directing them to the proper cellular location. In this sense, the Golgi apparatus functions as a kind of post office for many of the molecules produced by the cell. The molecules produced by some viruses interfere with this process in the case of MHC-1 molecules.

Yet other molecules produced by some viruses can interfere with the synthesis of MHC-1 proteins, leading to a decrease in their expression on the membrane. Other viral molecules allow MHC-1 molecules to reach the membrane with their peptides, but accelerate the normal process of recycling these molecules, stimulating the process of endocytosis by which the cells incorporate molecules from the outside. The MHC-1 molecules are thus rapidly removed from the membrane without allowing the T lymphocytes enough time to detect them.

However, TAP transporters are the most important target affected by immunoevasins of viruses. These proteins impede the activity of these transporters and prevent peptides from entering the endoplasmic reticulum. Under these conditions it is obviously impossible to load peptides onto MHC-1 molecules, and we already know that these molecules without their peptide are unstable and rapidly degraded. Four inhibitory proteins of the TAP transporters have been identified. These proteins prevent by different means that these transporters capture the peptides in the cytoplasm and introduce them inside the endoplasmic reticulum.

In addition to the mechanisms that block the expression of MHC-1 molecules and the presentation of peptides by these, some viruses also put into action mechanisms that hinder or prevent the expression of peptides in MHC-2 molecules. It should be recalled that one of the immune mechanisms that will be diminished without sufficient expression of peptides by MHC-2 molecules will be the production of the appropriate classes of high-affinity antibodies that can adequately neutralize viruses. Neutralizing antibodies are essential to eradicate viral infections, since activated CD8 T lymphocytes cannot eradicate the virus by themselves. Therefore, preventing effective antibodies from being produced at an adequate concentration is a good viral strategy to maintain the infection in a host.

Of course, affecting the peptide presentation process is not the only possible strategy to prevent virus-derived peptides from being presented to T cells. Another strategy is to prevent dendritic cells from maturing and becoming activated and traveling to the lymph nodes to perform their antigen-presenting and T-cell activating function. Indeed, some viruses have genes that interfere with the activation and maturation of dendritic cells. One such virus is the hepatitis C virus. Another one is the measles virus. Yet another defensive strategy employed by viruses is to affect the cell adhesion molecules needed for CD8 T cells or NK cells to bind to and kill infected cells. These latter cells will be discussed in some detail below.

## 7.4.- Natural Killer cells

Viruses have at their disposal many evasive mechanisms, but the immune system of animals has not stood still against the strategies used by these microorganisms to stay alive and continue their infectious capacity. Their evolution has also continued and has generated new defense mechanisms. One of them is the activity of the Natural Killer cells.

We have seen that one of the viral defense mechanisms against the immune system is to affect the expression of MHC-1 genes in infected cells. These molecules are responsible for binding peptides of the proteins synthesized by the cell itself and thus indicate whether the synthesized proteins are self-proteins or come from a foreign organism reproducing inside it. If MHC-1 molecules are not produced, the cell cannot express its identity, as it will lack "faces" to do it. In these situation, cytotoxic CD8 T cells, even if they have been correctly activated by cDC1 dendritic cells **(section 2.7)** that have captured the virus from the exterior, without being infected, and have presented peptides derived from viral proteins by cross-presentation, will not be able to attack the infected cell nor induce its apoptosis, as they cannot detect with their "mask", their T cell receptor, that the cell is infected. Therefore, viruses capable of preventing the expression of MHC-1 molecules in infected cells have a great advantage.

However, as we have seen, throughout evolution there has been, and continues to be, a true war between parasites and parasitized organisms. Each "technological advance" achieved by the former to improve their attack usually produces another "technological advance" in the parasitized or infected organisms that counteracts the former and improves their defense against them. In the case of viruses, the technological advance resulted in the generation of cells specialized in detecting the absence of “faces”, or even a decrease in its normal number, normally expressed by a cell of the organism. These specialized cells are the Natural Killer (NK) cells.

NK cells were identified because they are somewhat larger than T and B lymphocytes and possess distinctive cytoplasmic granules that contain cytotoxic proteins like those of activated CD8 T lymphocytes, including perforin and granzyme. As a result, they can kill tumor cells of certain types and cells infected by viruses without prior immunization. This property gave them their name, because CD8 T lymphocytes must be “trained” and activated to kill, while NK cells possess this ability “naturally”. Most of the body's NK cells are found in the liver, which is an important organ for the defense because it generates the complement proteins as well as the acute phase proteins. The reason for the abundance of these cells in the liver is not known with certainty. It is believed that these cells may participate, in addition to defensive functions, in organ homeostasis, but this has not been proven.

The absence or decrease in the number of MHC-1 molecules on the cell membrane reveals that the cell is not in a normal state. This abnormality may be caused by an infecting virus that manipulates the level of expression of MHC-1 genes, but it may also be that the cell has been transformed into a tumor cell. Tumor cells have mutated some of their genes and therefore show abnormal peptides in their MHC-1 molecules. This leads to their elimination by CD8 T cells and generates a selection pressure also in the tumor cells. Those expressing lower levels of MHC-1 molecules may escape death induced by CD8 T cells and will eventually prevail in the tumor, which will grow out of control. Therefore, we now see that in cases where some cells of the organism have stopped being team mates with the others, which occurs when cells are infected by a parasite and "seduced" into obeying it in pursuit of its reproductive interests, or when the cells have become malignant and seek to reproduce unchecked without obeying the orders of the rest of the organism telling them not to reproduce, the cells find it advantageous to decrease the expression of MHC-1 molecules, because in this way they avoid being eliminated by cytotoxic CD8 T cells.

This is when the NK cells come in. NK cells are patrolling the various organs and tissues in search of cells that try to hide from the activity of cytotoxic CD8 T cells by preventing the expression of MHC-1 molecules. NK cells are of the lymphoid lineage, that is, they are related to lymphocytes, not so much to granulocytes, but their functioning is more related to innate than to adaptive immunity, because they lack specific receptors to detect antigens and they are functional without having been previously stimulated by an antigen. **On the contrary, NK cells have specific receptors for self-molecules, including MHC-1 molecules**. When these receptors detect MHC-1 molecules they send a biochemical signal to the interior of the cell that prevents its killer activity. However, if an NK cell in contact with another cell does not receive that inhibiting signal then it secretes cytotoxic granules containing granzyme and perforin, as do CD8 T cells. As with CD8 T cells, perforin forms pores in the target cell membrane through which granzyme penetrates, inducing apoptosis. Alternatively, the NK cell can induce target cell death simply by contact, stimulating "**death receptors**" on the surface of the target cell. These death receptors, when stimulated by a specific molecule, trigger a biochemical mechanism within the cell that induces death by apoptosis. NK cells express on their surface the **TRAIL** molecule, which is a ligand for the death receptors **DR4** and **DR5**.

Besides their role in detecting the decline or absence of MHC-1 molecules, NK cells also participate in adaptive immunity in conjunction with antibodies. Antibodies of the IgG class capable of binding on the surface of our own cells to viral or modified self-antigens, which would indicate infection or tumor transformation, respectively, can be detected by a specific Fc receptor (the receptor called **FcγRIII**) expressed by NK cells. When this happens, NK cells also release their cytotoxic granules and induce death by apoptosis of the target cells. This process is called **antibody-mediated cell cytotoxicity** (ADCC) and is primarily involved in the control of viral infections. This mechanism requires activation of adaptive immunity to enhance the protective activity of innate immunity cells, such as NK cells, thus connecting these two types of immunity.

The activity of NK cells can be greatly stimulated by molecules and cytokines secreted by cells that have been infected by viruses. These molecules include interferons α and β (IFN-α and IFN-β) **(section 7.5)**, or IL-12, a cytokine secreted after pathogens are detected by dendritic cells and macrophages. NK cells activated by these molecules display 20 to 100 times the activity shown by NK cells not activated by these substances. This increased activity, particularly against virus-infected cells, provides time for adaptive immunity to generate cytotoxic CD8 T cells, which will be even more effective than NK cells in killing virus-infected cells, and also provides time for the generation of virus-neutralizing antibodies, which, as we have mentioned, are essential to control the infection along with the action of cytotoxic CD8 T cells. While the activity of NK cells is increased by the action of interferons and IL-12, these cells secrete high amounts of Interferon-γ (IFN-γ), an important cytokine that activates macrophages and dendritic cells. In addition to increasing the cytotoxic activity of macrophages, IFN-γ stimulates dendritic cells to present antigens and thus stimulates naïve CD4 T cells to differentiate towards the $T_H1$ phenotype, which also produces IFN-γ. As we can see, once the activation of NK cells is triggered, they accelerate the initiation of adaptive immunity and stimulate the activation of such important innate immunity cells as dendritic cells and macrophages.

The study of NK cells has led to the discovery of interesting facts about them, particularly how they "decide" whether they have found a healthy or a sick cell, in which case they must eliminate it. What the studies have revealed is that detecting healthy or sick cells and deciding to act on them is even more complicated for NK than for T cells, because it depends on a complicated balance between the action of activating receptors and the action of inhibiting receptors expressed on the surface of the NK cells, which does not only involve MHC-1 molecules. These receptors receive signals from the cells with which the NK cells interact. If the latter have suitable molecules on their surface capable of acting on the inhibitory receptors, these will inhibit the cytotoxic activity of the NK cell and the cell will not act by killing them. These molecules are usually abundantly expressed by healthy cells and among them are MHC-1 molecules, although these are not the only ones. Conversely, if a cell with which the NK cell interacts is not able to molecularly appease the natural aggressiveness of the NK cell, the NK cell will kill it. In addition, there are situations where cells in the body may express higher than normal levels of cytotoxicity-activating NK cell molecules. This happens, for example, when a cell has been infected or is under metabolic stress, as in the case of a tumor cell, or has suffered DNA damage that it has not been able to repair and that prevents it from functioning properly. This change in cell identity under stress is another important phenomenon detected by NK cells linked to the decrease in MHC-1 expression mentioned above.

Finally, it is interesting to mention that although they are part of innate immunity, NK cells undergo a process called **NK cell education**. The molecular details of this phenomenon are still being studied at the time of writing this (June 2020), but the result of this education is that various NK cells are adapted to detect different levels of expression of self-molecules before they act. In other words, different NK cells are calibrated in a particular way to act only when they detect particular levels of activating and inhibiting receptors on their target cells. The complexity of our immune system and its sophistication is phenomenal.

## 7.5.- Plasmacytoid dendritic cells and type I IFNs

Defense against viruses is undoubtedly fundamental to the survival of all organisms. Perhaps for this reason, as well as for the "cunning" displayed by viruses to escape the action of the immune system, animals have many mechanisms of action against them. In addition to cytotoxic CD8 T cells and NK cells, there are still additional mechanisms for slowing down the progression of viruses. One of these is carried out by a type of dendritic cells that we have not yet found: **plasmacytoid dendritic cells**. These cells are not found in the lymph nodes. They circulate in the blood and are disseminated throughout the various tissues, where they are awaiting potential attempts of viral infections. When these cells detect components of the viral particles, especially through their TLR-7 and TLR-9 Toll-like receptors, which respond to the presence of foreign nucleic acids, such as those characteristic of the viral particles that may have been captured by these cells, they produce and secrete very high quantities of molecules called **type I interferons**. These are mainly of two types: interferon-α (IFN-α) and interferon-β (IFN-β). There are thirteen different types of IFN-α, each produced by a different gene. These are the interferons produced mainly by plasmacytoid dendritic cells. There are only two types of IFN-β, which are not produced by plasmacytoid dendritic cells, but mainly by skin and connective tissue fibroblasts, which are responsible for the integrity of epithelial barriers and are very important in wound healing, and that also participate in the defense against viruses. However, both types of interferons can also be produced, although in smaller quantities, by T and B lymphocytes, macrophages, NK cells and even endothelial cells.

By secreting interferons, the cells that have directly captured the virus components communicate to other cells the information that there is a danger of virus infection. Once they have been detected by specific receptors on the outer membrane of the cells, interferons stimulate in them numerous anti-viral effects. These specific receptors activate and set in motion several molecular mechanisms inside the cells and can also trigger the production and secretion of other substances that activate adaptive immunity.

The effects exerted by interferons are of various types and complement each other to set up a comprehensive defense against viral infections. Firstly, they set in motion genes that direct the production of enzymes to degrade the viral components, in particular nucleic acids. Secondly, they activate the functioning of genes that will slow down or prevent protein synthesis inside the cells, and thus prevent the generation of viral proteins, which are necessary to produce new viral particles. Thirdly, interferons stimulate proteasome activity and the production of MHC-1 molecules, thus facilitating the presentation of peptides derived from viral proteins, which can help the infected cell to be detected by cytotoxic T cells. Finally, interferons can stimulate the generation of fever by binding to opiate receptors present in the neurons of the hypothalamus and, in addition, stimulate the production of prostaglandin E2, thus accelerating the adaptive immune response.

## 8.- COVID-19 AND YOUR IMMUNE SYSTEM

At the end of 2019, it is believed that on a date close to November 17, although the virus could have been already circulating for a few months, the first cases of a respiratory disease similar to the one that emerged in 2002 appeared in the Chinese city of Wuhan, capital of the province of Hubei, also in China, but in this case in the province of Guangdong, about 1,000 kilometers (around 620 miles) away from Wuhan. This last disease was called SARS (Severe Acute Respiratory Syndrome), and caused 8,422 reported cases worldwide, with a total of 774 deaths. The SARS outbreak appeared on November 27, 2002, and the epidemic was officially declared terminated by the World Health Organization on July 5, 2003.

The virus that caused the SARS epidemic (now called SARS-CoV-1) was identified as belonging to the family of coronaviruses, named for its characteristic protein "corona" that appears visible under the electron microscope and is slightly reminiscent of the solar corona that can be seen in total solar eclipses. The genome of this virus, which like those of other coronaviruses is very large, about 30,000 nucleotides, was sequenced in April 2003. Shortly thereafter, it was confirmed to be the cause of the disease by conducting experiments on macaque monkeys, which when infected by the virus developed a human-like disease. The virus genome indicated that its probable origin was some animal that was in frequent contact with humans. A month later, researchers discovered that the virus was found in a species of civet, more specifically in the so-called palm civet, although it did not appear to cause disease in these animals. This led to the elimination of more than 10,000 specimens of this animal in an attempt to control possible additional infections in humans. However, shortly afterwards it was found that the SARS-CoV-1 virus also infected other animals, such as a species of fox and an Asian weasel, and also domestic cats. In 2005, a study identified Asian bats as the animals that possibly acted as reservoirs of coronaviruses in the animal kingdom. These animals are chronically infected with numerous species of coronavirus, but these do not appear to cause disease in them. Further studies indicated that the virus had been able to pass from a bat species to the palm civet and from this species to infect humans. In 2017, a research group at the Wuhan

Institute of Virology was able to identify the bat species from which the SARS-CoV-1 virus originated. These bats were found in a cave in Yunnan Province, surprisingly more than 1,000 kilometers from Guangdong Province, where the epidemic broke out. This species was infected with a strain of coronavirus similar enough to the SARS-CoV-1 virus strain to conclude that it originated there. Because of their disturbing discovery, the researchers warned that an epidemic similar to that caused by SARS-CoV-1 could arise at any time if measures were not taken to limit contact between humans and animals in close contact with the bats.

These studies made it clear that the SARS-CoV-1 virus had emerged in a mutation process that some bat coronaviruses had undergone and that had allowed them to overcome the molecular and biological barrier and spillover to the human species and cause a zoonosis, that is, an infectious disease caused by a microorganism belonging to an animal species that does not normally infect humans. Throughout recent history, there have been several epidemics or pandemics caused by zoonoses. Perhaps the best known is the AIDS epidemic, caused by the HIV virus **(section 7.1)**, but there have been more recent epidemics, such as the 2009 swine flu epidemic. Epidemics generated by spillover infections are not limited to those suffered by humans, and other species can suffer from them as well. But what is the biological and molecular interspecies barrier, and what does it mean if a microorganism is able to go over it? To understand it better, we are going to have to deviate slightly from the immune system and go briefly into the also fascinating world of genetics and cell biology.

## 8.1.- Overcoming the Interspecies Barrier

The interspecies barrier is actually a genetic and molecular barrier generated by the genetic and molecular differences existing between different species and which generally limits their reproductive capabilities to only individuals of a given species. In other words, in general terms, individuals of a species can only reproduce with individuals of the same species, and not with individuals of other, even closely related, species. There are exceptions to this rule, as revealed by the fact that the human species possesses a certain percentage of the Neanderthal DNA in its genome, indicating that hybrids between individuals of both species were viable and could even reproduce with subjects of either species.

Why does this happen? Firstly, in each reproductive event, mutations occur, changes in the genes of the new member of the species with respect to their predecessors. These changes are not significant enough for the pieces of the cellular machinery generated from those genetic instructions to no longer fit properly with the pieces generated from genetic instructions from another individual of the opposite sex of the same generation. In other words, genetic changes that occur in only one generation are not important enough to prevent reproduction among members of the same generation of that species. However, if two populations of individuals of the same species reproduce in isolation for a sufficiently large number of generations, both populations will drift genetically from each other generation after generation. The cellular nuts and bolts generated from the instructions stored in the genome of both populations will gradually become different, generation after generation. If this situation continues for a long enough time, at least one or more of the parts of the cellular machinery generated from the instructions in the genome of one population will no longer fit properly with the parts of the cellular machinery generated by the genetic instructions of the other population. It is even possible that genes and cell parts have been lost or gained in one or other population during isolation. This incompatibility caused by the accumulation of mutations and genomic changes over time will prevent the functioning of cells derived from the fusion of two gametes from opposite sex individuals in each of the populations. At this point, we can say that two different species have been generated, genetically isolated from each other due to the incompatibility of their genomes to generate, together, a new viable organism or, at least, a new organism capable of reproducing itself.

Certainly, there may be other incompatibilities generated by genetic changes which, however, do not prevent the functioning of hybrid cells, although they do prevent cross-reproduction. For example, it is not easy to imagine how a Chihuahua and a Great Dane could cross reproduce. The difference in size between the two dog breeds makes cross breeding by natural means practically impossible. However, both breeds could be reproduced by artificial means, although the bitch should probably always be the larger breed, because the smaller breed would not be able to sustain the pregnancy. However, this situation has been created by the rapid artificial selection carried out with the different dog breeds and is not at all common in Nature.

Exceptions aside, we can now see a little more clearly what the interspecies barrier entails. It is an incompatibility between genes and proteins generated by the cells of one or other species that prevents their cross reproduction. Well, this molecular incompatibility also has a fundamental effect on viruses, which are generally capable of infecting one or a few genetically related species but are not capable of infecting species that are genetically more distant. As with the molecular nuts and bolts of the cells of two different species, the molecular parts that belong to the viruses have had to adapt to be compatible with the molecular parts of one or a few related species. This adaptation generally makes it impossible for a virus that infects one or a few related species to infect other species that are genetically distant from it.

However, it is a fact that viruses whose genome does not consist of DNA but of RNA, a group of viruses to which the AIDS, influenza and coronavirus viruses belong, can mutate with some frequency. This frequent mutation means that, as a virus infects one or a few cells of an animal or human being and this generates more and more viral particles, in the end the infected individual is infected not with a single virus but with a population of viruses, a "cloud" formed by up to thousands of mutant virus variants that differ more or less from the original. This population of mutant viruses, derived from the original one, has been called a **quasispecies**. To better understand this concept, we can assimilate the genome of a virus to a phrase, which also consists of a succession of letters and spaces. Let's assume the phrase: "I won't call my father an idiot anymore". Let us also suppose that, as in the past, we punish a child who has called his father an idiot by writing the phrase a hundred times on a piece of paper. Being rebellious as he is, the child decides to make a mistake with every copy of the sentence. So, he begins by writing: "I won’t call my mother an idiot anymore". And then he writes, "He won’t call my father an idiot anymore." And then he writes, "I won’t call my father an ideot anymore." And he continues to generate sentences with errors, until he completes a hundred. At the end of his task, the child will have generated a hundred slightly different phrases that will constitute a quasispecies of the original phrase: "I won't call my father an idiot anymore".

The different variants of the viruses that form a quasispecies are each located at one point or another in the so-called **adaptive landscape**. This landscape is like the natural landscape that we can visualize if we imagine a mountain range. There are peaks and valleys in that mountain range. The highest peaks of the adaptive landscape would be occupied by mutants that are very efficient at reproducing in the cells of the species that the virus infects. The lower peaks or valleys would be occupied by virus variants that are less efficient at reproducing in those cells. Since each virus variant, when it reproduces inside a cell, generates a small population of mutant viruses, not identical clones of the original virus, when the viruses have reproduced in hundreds or thousands of cells, the population of virus variants that form the quasispecies in the infected individual is practically in equilibrium, i.e. generation after generation of viruses, the population of these will have a similar distribution of the generated virus variants, although sometimes there may be deviations from this equilibrium point. In any case, this population will be dominated by the variants that are most effective in reproducing in the cells of the species that the virus infects, but there will also be mutants who, although not the most effective in infecting the cells of that species, may be able to infect the cells of other species more effectively, if they come into contact with them.

We can imagine that different animal species living more or less closely together are situated along a line at different distances from each other, distances that represent their susceptibility to be infected by a virus that optimally infects a particular species. This line places a species at the evolutionary (phylogenetic) distance it has from the others it may encounter. For example, if the line had a distance in units from 1 to 100, and we placed the human species arbitrarily at number 90, chimpanzees could be placed at number 89, orangutans at 85, bats at 72, mice at 70, and mushrooms at 10, to give an example. Although the distances chosen intend to reflect an approximate genetic distance between the different species mentioned and the human species, they are not exact distances that accurately reflect the genetic relationship between them.

In any case, what the line tries to reflect is that it is much more likely that a virus infecting the chimpanzee can "jump" to infect a human or an orangutan than a mouse. Obviously in the first two cases the distance of the "jump" is much less than in the last.

On the previous line we can place the distribution of variants of a virus quasispecies according to their ability to infect a given species. Each of the variants of a virus quasispecies capable of infecting a given species will be at a certain frequency in the virus population. This frequency will depend on its ability to infect the cells of that species. Therefore, the variants best adapted to infect the cells of that species will be more frequent and will coincide on the previous line with the number given to the species that the virus infects. The less infectious variants will be placed somewhat to the right or somewhat to the left of that number, following a distribution like that of the well-known bell curve. This means that in the distribution of virus variants infecting a given species, some will have an easier time infecting a closely related species than others; however, the amount of virus from those variants in the virus population infecting individuals of its canonical host species will be small, and it will be all the smaller the worse the variants are able to infect their host species, i.e. the further their infectivity is away from the optimal number corresponding to their species. The following figure illustrates this idea.

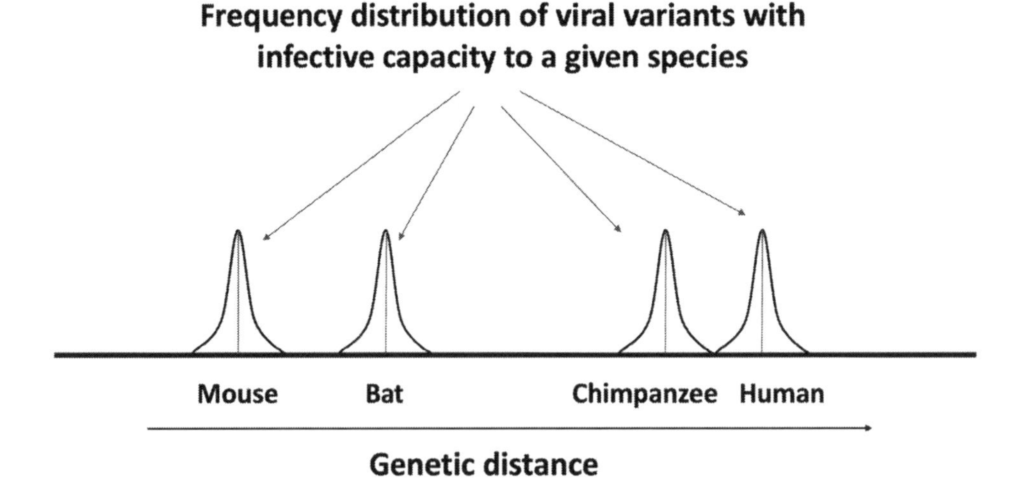

This implies that it is very unlikely that a virus of one species will infect another. This is so because many individuals of one species need to be infected in order for the population of virus infecting that species at that time to have a sufficient number of infected individuals with variants sufficiently different from the most effective, and majoritarian, to be able to infect an individual of another species. It is also necessary that at the moment when these variants have been generated and during the short time in which they are going to be viable, because they will die soon given that they cannot reproduce effectively in their canonical species, an infectious encounter (that is to say, it has to be proximity and even contact) happen between an individual of a species genetically close to the infected species and an individual of the infected species with the variants most capable of infecting that species.

Of course, what was mentioned above is only a simplification of reality with an educational objective, to make it easier to explain the idea of what can happen. What really happens is more complex, and that is for at least two reasons. The first is that not all individuals of a species are genetically concentrated in the same thin line, as shown in the figure. The individuals of a species are also distributed in a Gaussian bell according to their genetic susceptibility to be infected by one or another variant of the viral population that is generated when a virus infects a species. This increases the spread, i.e. the width of the curve of viral variants capable of infecting one species, and brings the species somewhat closer together in their susceptibility to infection by a virus from another species. In other words, there are chimpanzees more susceptible to be infected by a human virus than others, and there are humans more susceptible to be infected by a chimpanzee virus than others, not all are equally susceptible as represented in a simplified way by the line of the figure that divides each curve in two symmetrical parts. The second reason is that viruses not only produce variants with different infective capabilities on a continuous basis. This is what would be called antigenic drift. However, viruses, especially RNA viruses such as coronaviruses and influenza viruses, can also undergo the process of antigenic shift **(section 7.2)** through gene recombination between two virus variants that infect the same cell. This recombination can generate, even with very low probability, variants far from the optimal infectivity of the species it infects, which may however be able to infect other nearby species with effectiveness.

The direct spillover between genetically distant species is so unlikely that it does not normally occur. For a virus from a species far removed from the human species, such as any bat species, to be able to infect the human species it normally requires first spilling over to a species that is genetically closer to us. This is believed to have been the case with the coronavirus that caused the SARS outbreak in 2002, which passed from bats to civets and other animals that are genetically closer to humans. At the time of writing (April 2020, during the Global Confinement), it is not known from which species the SARS-CoV-2 virus could have spread. However, studies indicate that the bat coronavirus RaTG13, which infects the species *Rhinolophus affinis* has a genome that is 96.2% identical to that of SARS-CoV-2, although the protein that it uses to bind to the receptor that allows it to infect cells is quite different from that of the bat virus, which in principle makes it incapable of infecting human cells. However, as we have explained above, it is possible that some rare mutants or recombinants bear mutations in that protein allowing them to bind to the receptor and this has made possible the direct spillover from that bat species to the human species. We will have to wait for future studies to find out what the precise origin of this virus is.

## 8.2.- COVID-19

COVID-19 (coronavirus disease 2019), so named by the World Health Organization, is the disease caused by infection with the SARS-Cov-2 virus. This virus, like SARS-CoV-1, infects cells that have on their surface a particular protein to which the virus can bind strongly, thus triggering a molecular process that leads to the fusion of the virus membrane with the cytoplasmic cell membrane and the internalization of the virus by the cell. This specific protein is called ACE2, or angiotensin-converting enzyme 2. This protein is expressed by cells of the lung, arteries, intestine, heart and kidney, so potentially SARS-CoV-2 virus could infect any of these cell types and cause disease. The ACE2 enzyme is very important for regulating blood pressure because it converts the hormone angiotensin II, a peptide hormone, into angiotensin 1-7. While angiotensin II is a vasoconstrictor, and therefore increases blood pressure, angiotensin 1-7 is a vasodilator and lowers it. The correct expression and activity of ACE2 is therefore essential for the control of arterial pressure. However, the SARS-CoV-2 virus has nothing to do with blood pressure. Simply, this virus, as well as other coronaviruses, throughout evolution have been able to develop infectious mechanisms that use a protein necessary for the life of the animals that these can neither eliminate nor mutate in a significant way to prevent infection by coronaviruses without self-causing serious damage.

The symptoms of COVID-19 disease, as well as its consequences, are highly variable. The most common symptoms include fever, dry cough, and shortness of breath. Other, more infrequent, symptoms include throat irritation, diarrhea, muscle and/or abdominal pain, and sputum production. Transmission from one person to another usually occurs by inhaling small droplets containing the virus that are produced by infected people when they cough, sneeze, speak or sing. The smaller droplets produced can float in the air and if the air is static, i.e. there is no draught, the droplets can remain suspended for several hours, as if in a deadly, invisible fog, and could be inhaled by those who enter or are in that closed, poorly ventilated space. Larger droplets fall more quickly and can be deposited on the surfaces of objects. The virus can remain on them for up to 72 hours. By touching the contaminated surface with the hands and then touching the mouth, nose or eyes, the virus can also infect the epithelial cells of the nose (even through the nasolacrimal duct) or throat. Obviously, closer contact between people, such as kissing or hugging, can also lead to infection. After the initial infection, an asymptomatic incubation period begins, which can last from 2 to 14 days, although the average period is 5 days.

Not all people infected with the virus develop symptoms. Many do not. Others develop mild symptoms, such as a slight fever. Still others may develop a higher fever and various aches and pains. Finally, others may develop severe pneumonia. People who develop a more severe illness include mostly older people. Mortality is therefore higher the older the person is.

There is still a lot to learn from this disease to explain all the characteristics that it displays and even others that we do not mention. As far as we are concerned, we are going to use the properties of the disease to try to learn something more about how the immune system can protect us or not, and in the latter case, how it can even kill us. To do this, we will start by explaining the curious relationship between bats, from which SARS-CoV-2 and other coronaviruses that have affected the human species derive, and the immune system of these friendly little animals.

## 8.3.- The curious relationship between bats and coronaviruses

In addition to the coronaviruses SARS-CoV-1 and SARS-CoV-2 there have been recent outbreaks of other coronaviruses that, fortunately, have not caused as much pain as the last one. In 2011, a new coronavirus appeared in the Middle East, causing a respiratory syndrome similar to SARS, although not in all infected people. The virus was called MERS-CoV (Middle-East Respiratory Syndrome) and, as far as is known, appeared in the Arabian Peninsula. Unlike SARS-CoV-1, this virus is still in circulation, but fortunately the epidemic what brought under control thanks to its low infectivity rate, as each infected person infected only one other or none on average.

In 2016, another new coronavirus, which appeared in China's Guandong Province, where SARS-CoV-1 also appeared, caused a severe epidemic in farm pigs, which died of severe diarrhea. This virus, called SADS (Severe Acute Diarrhea Syndrome), killed 20,000 young pigs, and reached a mortality rate of 90% in those animals. Fortunately, it didn't affect humans this time.

Thus, the new SARS-CoV-2 coronavirus is but one of a series of new coronaviruses that have appeared in recent years. Given this observation, we can ask ourselves some questions: Why have new viruses appeared in recent years? Why do new viruses appear in China or Asia? Data obtained, mainly by Chinese researchers, since the appearance of the first SARS epidemic until now, have revealed that most of the known species of coronavirus, of which seven are capable of infecting humans, are found in bats in China, animals that these viruses chronically infect without causing them excessive damage. From these animals, coronaviruses can spillover to other species and affect humans and domestic animals. Therefore, we can also ask ourselves: What is so special about bats that they can host coronaviruses without being seriously affected?

Bats are not only infected by coronavirus, although it is estimated that this group of viruses is probably the one that infects them the most. Bats can also be vehicles for other RNA viruses, such as rabies, Zika and Ebola viruses. Bats appear to be a class of animals that can be infected by numerous viruses and yet not cause them serious epidemics or illness.

This property makes bats the most important reservoir of viruses, especially of coronavirus, in Nature. This is because bats are the second most numerous order of mammals on the planet, after rodents. There are more than 1,300 species of bats, 20% of all mammal species. Most of them live in China and South Asia. This information alone helps explain why new coronaviruses infecting humans are emerging in Asia and particularly in China. The great diversity of coronavirus species that infect bats also increases the probability of recombination between two different viruses and the generation of a new recombinant virus, a mixture of two others, which can be infectious and for which we would lack immune defense. However, most viruses generated by recombination will probably not be effective in infecting bat cells and will die soon. A few, however, can survive and could sooner or later spillover to another species, including humans.

As we know, bats are the only flying mammals, which allows them to travel greater distances than other mammals and spread over wide territories more easily. This maximizes the likelihood of infectious encounters with numerous other species, including, of course, humans, who have also reduced much of the habitat of these animals due to their unstoppable and senseless expansion around the planet, causing intense deforestation and significant effects on ecosystems. Without a doubt, one of the countries where the expansion of the human species has been most rapid in recent decades, both in terms of the number of individuals and the intensity of resources consumed and the effects on the environment, has been China.

But what makes bats so susceptible to harboring coronaviruses and other virus species? To understand this, we should briefly dive into the immune system of these animals, where the explanation can be found.

To begin with, it must be said that with more than 1,300 species of bats, it is not expected that they all possess identical immune systems, nor that the mechanisms used by the immune systems of bats to be able to live with the viruses and control them without causing disease are the same. However, studies in several bat species have revealed some surprising adaptations of the immune system of these animals to the viruses.

Firstly, sequencing of the genomes of some bat species indicates that they have a lower percentage of genes dedicated to immune functions. While humans or mice have 7% of their genes dedicated to the functioning of the immune system, the bat species whose genomes have been sequenced have only 3% to 3.5% of their genes dedicated to this system. However, it is not yet clear whether bats possess additional, yet undiscovered, genes that are dedicated to the functioning of the immune system.

Other studies have revealed that bat NK cells do not possess a set of receptors like those of other mammals. It should be remembered **(section 7.4)** that NK cells are a fundamental type of cell in the fight against viral infections and that they are equipped with a set of receptors that activate or inhibit their cytotoxic activity. We haven't mentioned this before, but NK cells have two sets of genes that inhibit their cytotoxic activity, capable of detecting MHC-1 molecules, which indicate that the cells are healthy. The first set is made up of genes called **KIR** (killer cell immunoglobulin-like receptors) and the second, **KLR** (killer cell lectin-like receptors), also known as **Ly49** receptors. Well, neither of these two sets of genes is found in the genomes of some bat species whose genome has been sequenced, indicating that these animals must possess a different set of receptors to detect MHC-1 molecules. However, other studies indicate that certain species of bats have expanded genes of the KLR family in their genomes that generate a balance in the activity of NK cells more inclined towards inhibition than activation, i.e. towards a greater tolerance of cells that may be infected by viruses, to which they do not easily induce apoptosis. However, the spread of these genes may also allow for greater finesse in detecting uninfected cells compared to those infected and manipulated by viruses to try to go undetected by NK cells. This would perhaps lead to greater effectiveness of these cells in controlling viral infections. In any case, a strong inflammatory response would be avoided. This inflammatory response depends on the signaling power of the activating receptors, which stimulate the production of cytokines and chemokines. Well, the NK receptors of many bats have a mode of signaling that is less powerful than the NK cell receptors of other mammals and the induction of cytokines and chemokines by these cells is less, which decreases the inflammatory response.

The potential superiority in the detection of cells expressing MHC-1, whose low levels of expression may indicate that the cells are infected, is accompanied in bat genomes by an expansion of the genes of the MHC-1 complex, some of which are located at different sites of the chromosomes from where the canonical locus that brings together the MHC-1 and MHC-2 genes is found in the rest of mammals. The increased number of MHC-1 genes may result in an increase in the amount of these molecules expressed on the surface of the cells, which again places the NK cells in an inhibitory mode.

In addition to an expansion of genes from the Major Histocompatibility Complex, the MHC-1 molecules themselves are different from those of other mammals. Firstly, bat MHC-1 molecules can accommodate in their binding cleft peptides of a longer length than other mammals, which possibly also increases the diversity of peptides that these molecules are able to present. The amino acids involved in the binding of peptides in the cleft of some of these molecules are also different from those present in other mammals, which may reflect an evolutionary process that has selected MHC-1 variants capable of presenting more efficiently peptides from the viruses that most frequently infect bats. Indeed, the study of the MHC-1 molecular structure in one bat species (*Pteropus alecto*) has revealed the presence of three additional amino acids not found in other mammal species. These three amino acids are involved in the formation of an additional ionic bond between the MHC-1 molecule and the peptides it presents, which is thought to be related to a greater efficacy for the presentation of virus-derived peptides and with the activation of a more effective response by T cells.

However, perhaps the greatest adaptation of the bat immune system has occurred in the type I interferon genes, whose antiviral actions we discussed in **section 7.5**. Bats may tolerate viral infections to a large extent by minimizing the pro-inflammatory effects of type I interferons. These are secreted by various cells, after detecting some viral components with their Toll-like receptors or other receptors of innate immunity. Viral nucleic acids are one of these components. For example, the single-stranded viral RNA that makes up the genome of RNA viruses can be detected by the TLR-7 receptor. TLR-9 detects viral or bacterial DNA instead. Be that as it may. These receptors, when stimulated by viral components, trigger a series of molecular events in the cytoplasm that lead to the transport to the nucleus of transcription factors that set in motion the genes for the production of type I interferons. These are secreted to the outside and are detected by neighboring cells, which may not have directly detected any viral component, but which thus know that their neighbors have. Once detected by means of receptors specific to them, these receptors in turn trigger a biochemical signal inside the cell that activate genes responsible for carrying out the antiviral defense mechanisms outlined in **section 7.5**. These genes are generically called **interferon-stimulated genes (ISGs)** and include some that will stimulate a strong inflammatory response as well as the generation of fever.

Different interferons interact with different receptors and induce a different set of ISGs, depending on the set of interferons presently stimulated by the signal received by TLRs or other receptors. This ultimately determines whether the effects on the body are more beneficial than harmful. This is because an exacerbated interferon response can lead to a strong inflammatory response that will cause significant collateral damage, while an insufficient response can ultimately leave us defenseless against the progression of the viral infection.

Studies of various bat species indicate that the interferon gene family has expanded in these animals. This expansion seems to allow the generation of more "personalized" responses to the infection with specific viruses. Most importantly, however, while this response can be very effective in preventing the progression of the viral infection, it does not trigger a powerful inflammatory response that could cause severe damage to the organs most affected by the infection, such as the lungs in the case of coronaviruses.

This increased diversity in the response of type I interferons in bats is thought to be related to their evolutionary history. The fact that bats are the only animals capable of flight doesn't come free to them. During the flight, the energy and metabolic needs are so high that the body temperature of these animals can reach 41°C. The high metabolic rate required to generate the energy to sustain flight and the high temperature generate oxygen radicals in their mitochondria that damage the mitochondrial DNA and these organelles, with the consequence that the DNA in the mitochondria can pass into the cytosol. The presence of DNA in the cytosol is also indicative of possible damage to the DNA in the nucleus and may be mistaken for an attempted viral infection. For these two reasons, bats, first, have very efficient DNA repair machinery. Secondly, and more importantly for the topic at hand, bat cells have decreased their ability to develop an interferon response to the presence of DNA in the cytoplasm, since this DNA can be self-DNA and reacting strongly against it would generate a harmful autoimmune attack.

One way in which many bat species have managed over the course of evolution to avoid triggering an intense interferon response to the presence of cytosolic DNA has been the loss of genes responsible for the production of proteins specifically involved in the detection of DNA in the cytosol and which bind their activation to that of TLR receptors to trigger an interferon response. These proteins belong to a family of proteins called PYHIN. The genes to produce these proteins have been removed from the genome of the bat species that have been studied until now. Another way in which bat cells avoid triggering a strong response to the presence of cytosolic DNA is the presence of a mutation in another cytosolic protein involved in DNA detection: the STING protein (stimulator of interferon genes). This protein functions as a direct sensor of cytosolic DNA and as an adaptor of the signal of type I interferons by various molecular mechanisms. The mutation that the STING gene possesses in bats diminishes their capacity of action and allows a more diminished interferon response against, in this case, DNA viruses. The intensity of the interferon response, however, seems to be sufficient to control the viruses.

Although these adaptations are intended to tolerate DNA in the cytoplasm, since the genes and mechanisms that detect this are in part common to those that detect viral RNA, the result is that the overall interferon response has been tempered in bat cells with respect to those of other mammals. In conclusion, all these adaptations make the bats able to tolerate a high load of infectious virus in their bodies without generating a strong inflammatory reaction that could cause their death. In other words, bats, due to their physiology and the metabolic demands of flight, have been forced to develop an immune system more tolerant of the damage generated in the cells themselves by the energy needs to sustain flight. This has led, probably, to the fact that they are also more tolerant to infections by various species of virus. The innate antiviral response is, however, capable of slowing down the progress of the virus through mechanisms that do not generate a strong inflammatory response, but which nevertheless give time to set in motion the mechanisms of adaptive immunity, which will eventually be able to eradicate the virus, even if this eradication takes longer than it would with more expeditious methods, but which would be more harmful.

### 8.4.- ¿WHY ARE SOME PEOPLE SEVERELY AFFECTED BY THE SARS-COV-2 VIRUS AND OTHERS HAVE NO SYMPTOMS?

Armed with the knowledge about the immune system and SARS-CoV-2 virus that we have learned so far, we can begin to look at possible causes why some people infected with SARS-CoV-2 are severely affected and many die, while others may pass the disease without ever developing noticeable symptoms. The answer to this question depends on several factors. We'll mention some of the more likely ones here, but it's almost certain that there are others still unknown.

A first factor that may influence the difference in severity of COVID-19 disease among people may come from differences in the ACE2 protein, which is that used by the virus to infect cells. If these differences exist, they could affect the affinity with which the virus binds to ACE2 on the surface of the cells. The SARS-CoV-2 virus uses a protein in its "corona" called the S protein (Spike) to bind to the ACE2 protein in cells and infect them. The S protein is made up of two subunits, S1 and S2, which perform different functions. S1 binds to the ACE2 protein and S2 is necessary for the fusion of the virus membrane with the cell membrane and the entry of the virus into the cell. ACE2 is also the protein used by SARS-CoV-1 virus. To date, no studies have indicated the presence of polymorphisms in the human ACE2 gene that could affect the infectious efficiency of the virus. These polymorphisms, if they exist, may not have a decisive effect on explaining the differences in the evolution of the disease observed in different people.

However, as we know, not only the affinity of the binding between two proteins is important to allow these to interact for long enough to perform their function. Avidity is also very important, sometimes even more so than affinity. Avidity, as we know, depends on the number of binding points, in this case the number of binding points that a virus particle can establish with the surface of a cell. For this reason, the level of expression of the ACE2 protein, i.e. the number of these molecules that can be located on the surface of the different cells that express this protein, could also be a determining factor affecting the entry of the virus into the cells. This level of expression is certainly not the same in different people and is most likely distributed in the population according to a normal curve, i.e. a Gaussian bell curve. If the amount of ACE2 protein on the surface of the cells needed to be high for the virus to cause an intense and effective infection, for example, if only 20% of the human population possessed a sufficient amount of ACE2, this could help explain why a similar percentage of people develop a more severe disease, which could be related to the higher infection efficiency of the viral particles in those people, which would favor the course of the infection. Indeed, people with lower ACE2 levels would be less effectively infected by the virus particles, giving the immune system more time to control the infection.

In addition to the ACE2 protein, the SARS-CoV-2 coronavirus requires the activity of another protein in the cell membrane. This protein is called TMPRSS2 and is an enzyme that cuts protein chains, that is, it is a protease, as are, remember, many proteins of the complement cascade. The activity of TMPRSS2 is necessary to cut the S2 protein of the virus once S1 has bound to ACE2 and thus generate a fragment of that protein, called a fusion peptide. This is a peptide capable of inserting itself into the cell membrane and facilitating the fusion between the virus membrane and the cell membrane. Therefore, the levels of expression of this protein could also influence the entry of viral particles into different cells. As in the previous case, the levels of expression of this protease will be distributed, most probably, according to a Gaussian curve and there will be people who will possess a greater number per cell of TMPRSS2 molecules than others and may thus be more susceptible to an initial infection and to the progression of the infection once it has begun.

The levels of expression of ACE2 and TMPRSS2 could therefore influence a factor that is fundamental to the control of all infections, as we discussed at the beginning: time. If most of the virus particles produced by a cell and released to the exterior can bind to and infect other neighboring cells, the infection will progress in that organism very quickly. However, if a smaller fraction of viral particles once released from an infected cell can bind to neighboring cells, because the levels of ACE2 or TMPRSS2 in the organism in which the virus is are not very high, then the viral infection will progress more slowly. This slower progression may allow enough time for the development of an adaptive immune response that controls and eventually eliminates the infection. Unfortunately, a rapid infection that cannot be controlled will already have caused serious damage to the cells on the surface of the bronchial and alveolar lungs before the adaptive immune response has had time to activate, which may trigger an infection by airborne bacteria that are also found lining the lung surface, but kept at bay by the integrity of the lung epithelial barrier. If the damage to this barrier caused by the virus is too great and occurs at a faster rate than it can be repaired, the bacteria will be able to penetrate that barrier and generate an intense inflammatory response. This intense response can lead to what are called **cytokine storms**, more academically called **cytokine release syndrome**. This syndrome is triggered when high numbers of monocytes, dendritic cells, macrophages, B and T lymphocytes, and NK cells are simultaneously activated and release a high amount of inflammatory cytokines, which in turn activate even more immune cells that will produce even more cytokines and further inflammation. The endothelial cells of the blood vessels are also important players in the production of cytokines that lead to the cytokine storm. The production of cytokines is also accompanied by a high production of chemokines that attract lymphocytes to the sites of inflammation. In the case of other viral infections affecting the lung, such as influenza, it has been found that the endothelial cells of the pulmonary blood vessels are primarily responsible for the production of the cytokines that cause the storm affecting that organ. The infiltration of lymphocytes and their activity on the blood vessels of the lung hinders the vital function of this organ, which is none other than the exchange of gases with the environment. If lung function is severely diminished, death by suffocation occurs. In some cases, sepsis can also be generated, which also leads to death. Clearly, in this situation, preventing or stopping the cytokine storm is vital, and for that we must intervene in some way to limit the activity of the immune system. Fortunately, not all cytokine storms are of the same severity, and like hurricanes, they are rated on a scale of 1 to 5. The administration of antihistamine and anti-inflammatory drugs and certain monoclonal antibodies against certain cytokines, such as IL-6, is useful, although drug treatment with these substances is not always able to stop the most serious storms.

How effectively the virus can spread its population and infect lung or other cells is not the only factor that can shorten or lengthen the time it takes for the adaptive immune system to control an infection. What we know about the immune system of bats in relation to coronaviruses indicates that there may be at least two other important factors.

The first factor is the intensity of the type I interferon response. This response, again, depends on the efficiency of the set of genes that each person has and will be distributed in the population according to a Gaussian curve. If the type I interferon response is adequate, the reproduction of the virus in the cells of the bronchial epithelium it infects will be slowed down and the virus will not be able to cause significant damage to that epithelium. This will give the adaptive immune system time to kick in before the damage is too great and the infection will be cleared before it could cause serious illness. It should be noted that this adaptive response not only involves the generation of neutralizing antibodies against the virus, which will prevent its attachment to the cells, but also involves the generation of cytotoxic CD8 T lymphocytes that will be able to kill the infected cells. If these are scarce because the interferon response has controlled the infection well and these cells are not much required, these lymphocytes will not cause excessive collateral damage.

If, however, the interferon response is not adequate, it is possible that the virus will reproduce more rapidly and damage the lung epithelium more severely. When the adaptive response is generated, it will not be able to stop the infection. Worse still, CD8 T lymphocytes will kill the probably numerous cells infected by the coronavirus and cause significant damage to the lung epithelium. In these circumstances, the bacteria that are inhaled with the air and those that cover the bronchial epithelium will be able to penetrate through this and trigger a strong inflammatory response by the many macrophages on the other side of the epithelium. These can trigger the above-mentioned cytokine storm, and will thus recruit numerous monocytes, macrophages, neutrophils and CD8 T cells into the virus-damaged tissue, greatly exacerbating the damage to the lung and significantly, even critically, compromising the function of this organ. In addition, macrophages will also secrete proteolytic enzymes such as metalloproteases **(section 2.5)**, which will damage surrounding tissue.

This provides us with a valuable lesson in understanding the limits of the immune system and why it can be deadly in its defensive activity if these limits are exceeded. It is clear that the adaptive response against the virus that involves the generation of cytotoxic CD8 T cells is effective when most cells in a tissue or organ are not infected and thanks to the innate response have been able to defend well against the attack of the virus, even if they have not eradicated it. If the innate immune response is not adequate to stop the virus, it extends widely over a vital organ and a widespread adaptive cytotoxic response is triggered, it may cause more harm than good.

The second factor that could affect the effectiveness of the immune response and the ability to stop virus progression is the type of MHC-1 and MHC-2 genes that each person has in their genome. We have seen before **(section 5.2.2.6)** that these genes are characterized by being polygenic and polymorphic. Polygeny involved the presence of multiple genes for MHC-1 and MHC-2 in each person's genome. Polymorphism meant that there were numerous forms of these genes in the human population. Combining polygenesis and polymorphism, the situation is generated that each person has practically a unique set of MHC-1 and MHC-2 genes, hardly present in other people. Each person will therefore have a set of MHC-1 and MHC-2 genes that will be more or less effective in terms of their ability to present peptides derived from SARS-CoV-2 coronavirus proteins. Depending on this efficiency, it is possible that the adaptive immune response, both the humoral response for the generation of antibodies of the right kind, and the cellular response, with the generation of CD8 T cells, may be more or less efficient and thus affect the development of the disease.

The above factors may help explain why people of the same age and similar general health conditions may nevertheless develop very different courses of the disease. Of course, the same considerations may be relevant to explain a different evolution of other infectious diseases and why some overcome them while others succumb to them. There are also other factors, such as the level of stress or physical activity, and in particular the general nutritional status, which are essential for the immune system to have the energy necessary to mount an effective response.

And the need for energy to mount an effective and rapid response, conveniently introduces us to the possible explanation as to why mortality from COVID-19 disease is higher the older the people suffering from it. The generation of energy in the form of ATP takes place mainly in the mitochondria. ATP is critical in providing energy for DNA reproduction during clonal expansion, as well as for the generation and secretion of cytokines and inflammatory response mediators.

One of the characteristics of aging, which is a process of genetic and molecular degeneration that impacts the function of cells, is that mitochondria lose efficiency in generating ATP. This means that all the cells of the immune system of older people will have lost effectiveness in generating the energy needed to mount an innate and adaptive response as quickly as possible; to secrete adequate amounts of interferon; to phosphorylate signaling proteins, which requires ATP; for gene expression, which requires ATP for the synthesis of messenger RNA and for protein synthesis, which also requires large amounts of ATP. Consequently, it will be more difficult, or even impossible, for older people to generate an immune response fast enough at the onset of infection to effectively contain the progression of the viral infection. The situation is comparable to that which would arise if people of different ages had to take shelter from a threat, perhaps from a dangerous animal attacking them, by running. Clearly, younger people, who can generate energy more efficiently, will run faster than older people in general and are more likely save themselves. In the same way, people's immune systems can "run" at higher or lower speeds. It is even possible that a young immune system be partially defective so that it has allowed the person to survive in a clean and hygienic environment, where vaccines are administered for the most dangerous diseases leaving the immune system prepared, trained, to face the most dangerous threats. However, that same immune system may not be effective enough to control infection with a new virus for which previous defenses are lacking.

The above considerations about the factors that may affect the efficiency of the immune system may also help to explain the significant differences in mortality due to COVID-19 that have been observed in different countries. It is true that adequate epidemiological and clinical decision-making to stop the pandemic in time can affect the extent of the epidemic in a given territory, but it cannot explain the different mortality rates observed. It is possible that these are mainly due to genetic factors, i.e. genetic differences between different human populations living in different regions of the planet. Differences in the alleles of the MHC genes, the genes of type I interferons, ACE2 and TMPRSS2 proteins, and even other genes not yet known but which could also participate in the effectiveness of the antiviral defense mechanisms of the immune system, could, together or separately, affect, as we have said, the severity of the disease caused by SARS-CoV-2 infection.

And with these words, we end this little tour of the world of immunology. I hope that you have been interested, enlightened and amazed, and that you can now appreciate a little better the extraordinary mechanisms that take place in your body, every day, to keep you alive in the face of the constant threats that are always lurking.

## 9.- Bibliography

To elaborate this book, I have consulted several bibliographical sources that I list below. Of particular importance to me have been the Immunology textbooks that I include in the corresponding section. It has also been very helpful to consult the English version of the Wikipedia encyclopedia, which sometimes has excellent entries that are very well documented and based on referenced scientific articles. Finally, I have also consulted specialized scientific publications which I include at the end of this section.

### 9.1.- Text Books


1. Janeway's Immunobiology (Ninth Edition, March 22, 2016), by Kenneth M. Murphy (Author), Casey Weaver (Author). W. W. Norton & Company Ed. ISBN-10: 0815345054; ISBN-13: 978-0815345053.

2. Case Studies in Immunology: A Clinical Companion (Seventh Edition, February 1, 2016), by Raif S. Geha (Author), Luigi Notarangelo (Author). W. W. Norton & Company Ed. ISBN-10: 9780815345121; ISBN-13: 978-0815345121.

3. Fundamental Immunology (Seventh Edition, December 19, 2012), by William E. Paul (Author). LWW Ed. ISBN-10: 9781451117837; ISBN-13: 978-1451117837.

4. Principles of Mucosal Immunology (First Edition, April 18, 2012) by Society for Mucosal Immunology (Author), Phillip D. Smith (Editor), Thomas T. MacDonald (Editor), Richard S. Blumberg (Editor). Garland Science, Ed. ISBN-10: 9780815344438; ISBN-13: 978-0815344438


### 9.2.- Scientific papers


Ahn, M., Anderson, D. E., Zhang, Q., Tan, C. W., Lim, B. L., Luko, K., et al. (2019). Dampened NLRP3-mediated inflammation in bats and implications for a special viral reservoir host. *Nature Microbiology*, *4*(5), 1–14. http://doi.org/10.1038/s41564-019-0371-3

Akondy, R. S., Fitch, M., Edupuganti, S., Yang, S., Kissick, H. T., Li, K. W., et al. (2017). Origin and differentiation of human memory CD8 T cells after vaccination. *Nature, 552*(7685), 362–367. http://doi.org/10.1038/nature24633

Albanese, M., Tagawa, T., Buschle, A., & Hammerschmidt, W. (2017). MicroRNAs of Epstein-Barr Virus Control Innate and Adaptive Antiviral Immunity. *Journal of Virology, 91*(16), 355. http://doi.org/10.1128/JVI.01667-16

Alcover, A., Alarcón, B., & Di Bartolo, V. (2018). Cell Biology of T Cell Receptor Expression and Regulation. *Annual Review of Immunology, 36*(1), 103–125. http://doi.org/10.1146/annurev-immunol-042617-053429

Allen, T. M., Brehm, M. A., Bridges, S., Ferguson, S., Kumar, P., Mirochnitchenko, O., et al. (2019). Humanized immune system mouse models: progress, challenges and opportunities. *Nature Immunology, 20*(7), 770–774. http://doi.org/10.1038/s41590-019-0416-z

Amsen, D., Helbig, C., & Backer, R. A. (2015). Notch in T Cell Differentiation: All Things Considered. *Trends in Immunology, 36*(12), 802–814. http://doi.org/10.1016/j.it.2015.10.007

Andersen, K. G., Rambaut, A., Lipkin, W. I., Holmes, E. C., & Garry, R. F. (2020). The proximal origin of SARS-CoV-2. *Nature Medicine, 26*(4), 450–452. http://doi.org/10.1038/s41591-020-0820-9

Banerjee, A., Baker, M. L., Kulcsar, K., Misra, V., Plowright, R., & Mossman, K. (2020). Novel Insights Into Immune Systems of Bats. *Frontiers in Immunology, 11*, 26. http://doi.org/10.3389/fimmu.2020.00026

Barral, D. C., & Brenner, M. B. (2007). CD1 antigen presentation: how it works. *Nature Reviews. Immunology, 7*(12), 929–941. http://doi.org/10.1038/nri2191

Barthels, C., Ogrinc, A., Steyer, V., Meier, S., Simon, F., Wimmer, M., et al. (2017). CD40-signalling abrogates induction of RORγt+ Treg cells by intestinal CD103+ DCs and causes fatal colitis. *Nature Communications, 8*(1), 14715–13. http://doi.org/10.1038/ncomms14715

Benn, C. S., Netea, M. G., Selin, L. K., & Aaby, P. (2013). A small jab - a big effect: nonspecific immunomodulation by vaccines. *Trends in Immunology, 34*(9), 431–439. http://doi.org/10.1016/j.it.2013.04.004

Biram, A., Davidzohn, N., & Shulman, Z. (2019). T cell interactions with B cells during germinal center formation, a three-step model. *Immunological Reviews, 288*(1), 37–48. http://doi.org/10.1111/imr.12737

Bordt, E. A., & Bilbo, S. D. (2020). Stressed-Out T Cells Fragment the Mind. *Trends in Immunology, 41*(2), 94–97. http://doi.org/10.1016/j.it.2019.12.008

Bosteels, C., Neyt, K., Vanheerswynghels, M., van Helden, M. J., Sichien, D., Debeuf, N., et al. (2020). Inflammatory Type 2 cDCs Acquire Features of cDC1s and Macrophages to Orchestrate Immunity to Respiratory Virus Infection. *Immunity, 52*(6), 1039–1056.e9. http://doi.org/10.1016/j.immuni.2020.04.005

Boudreau, J. E., & Hsu, K. C. (2018a). Natural Killer Cell Education and the Response to Infection and Cancer Therapy: Stay Tuned. *Trends in Immunology, 39*(3), 222–239. http://doi.org/10.1016/j.it.2017.12.001

Boudreau, J. E., & Hsu, K. C. (2018b). Natural killer cell education in human health and disease. *Current Opinion in Immunology, 50,* 102–111. http://doi.org/10.1016/j.coi.2017.11.003

Cartier, A., & Hla, T. (2019). Sphingosine 1-phosphate: Lipid signaling in pathology and therapy. *Science, 366*(6463), eaar5551. http://doi.org/10.1126/science.aar5551

Chang, V. T., Fernandes, R. A., Ganzinger, K. A., Lee, S. F., Siebold, C., McColl, J., et al. (2016). Initiation of T cell signaling by CD45 segregation at 'close contacts'. *Nature Immunology, 17*(5), 574–582. http://doi.org/10.1038/ni.3392

Chen, L., & Flies, D. B. (2013). Molecular mechanisms of T cell co-stimulation and co-inhibition. *Nature Reviews. Immunology, 13*(4), 227–242. http://doi.org/10.1038/nri3405

Chen, X., Liu, Q., & Xiang, A. P. (2018). CD8+CD28- T cells: not only age-related cells but a subset of regulatory T cells. *Cellular & Molecular Immunology, 15*(8), 734–736. http://doi.org/10.1038/cmi.2017.153

Chou, C., & Li, M. O. (2018). Re(de)fining Innate Lymphocyte Lineages in the Face of Cancer. *Cancer Immunology Research,* 6(4), 372–377. http://doi.org/10.1158/2326-6066.CIR-17-0440

Cohen, N. R., Garg, S., & Brenner, M. B. (2009). Antigen Presentation by CD1 Lipids, T Cells, and NKT Cells in Microbial Immunity. *Advances in Immunology, 102,* 1–94. http://doi.org/10.1016/S0065-2776(09)01201-2

Cruz-Muñoz, M. E., Valenzuela-Vázquez, L., Sánchez-Herrera, J., & Santa-Olalla Tapia, J. (2019). From the "missing self" hypothesis to adaptive NK cells: Insights of NK cell-mediated effector functions in immune surveillance. *Journal of Leukocyte Biology, 105*(5), 955–971. http://doi.org/10.1002/JLB.MR0618-224RR

Dilucca, M., Forcelloni, S., Georgakilas, A. G., Giansanti, A., & Pavlopoulou, A. (2020). Codon Usage and Phenotypic Divergences of SARS-CoV-2 Genes. *Viruses, 12*(5), 498. http://doi.org/10.3390/v12050498

Domingo, E., Sheldon, J., & Perales, C. (2012). Viral quasispecies evolution. *Microbiology and Molecular Biology Reviews : MMBR, 76*(2), 159–216. http://doi.org/10.1128/MMBR.05023-11

Duan, Z., Li, F.-Q., Wechsler, J., Meade-White, K., Williams, K., Benson, K. F., & Horwitz, M. (2004). A novel notch protein, N2N, targeted by neutrophil elastase and implicated in hereditary neutropenia. *Molecular and Cellular Biology*, *24*(1), 58–70. http://doi.org/10.1128/mcb.24.1.58-70.2004

Duffy, K. R., Wellard, C. J., Markham, J. F., Zhou, J. H. S., Holmberg, R., Hawkins, E. D., et al. (2012). Activation-induced B cell fates are selected by intracellular stochastic competition. *Science*, *335*(6066), 338–341. http://doi.org/10.1126/science.1213230

Fink, K. (2012). Origin and Function of Circulating Plasmablasts during Acute Viral Infections. *Frontiers in Immunology*, *3*, 78. http://doi.org/10.3389/fimmu.2012.00078

Ford, M. L. (2016). T Cell Cosignaling Molecules in Transplantation. *Immunity*, *44*(5), 1020–1033. http://doi.org/10.1016/j.immuni.2016.04.012

Freud, A. G., Mundy-Bosse, B. L., Yu, J., & Caligiuri, M. A. (2017). The Broad Spectrum of Human Natural Killer Cell Diversity. *Immunity*, *47*(5), 820–833. http://doi.org/10.1016/j.immuni.2017.10.008

Gascoigne, N. R. J., Rybakin, V., Acuto, O., & Brzostek, J. (2016). TCR Signal Strength and T Cell Development. *Annual Review of Cell and Developmental Biology*, *32*(1), 327–348. http://doi.org/10.1146/annurev-cellbio-111315-125324

Gennery, A. (2019). Recent advances in understanding RAG deficiencies. *F1000Research*, *8*(148), 148. http://doi.org/10.12688/f1000research.17056.1

Giamarellos-Bourboulis, E. J., Netea, M. G., Rovina, N., Akinosoglou, K., Antoniadou, A., Antonakos, N., et al. (2020). Complex Immune Dysregulation in COVID-19 Patients with Severe Respiratory Failure. *Cell Host & Microbe*, *27*(6), 992–1000.e3. http://doi.org/10.1016/j.chom.2020.04.009

Gravbrot, N., Gilbert-Gard, K., Mehta, P., Ghotmi, Y., Banerjee, M., Mazis, C., & Sundararajan, S. (2019). Therapeutic Monoclonal Antibodies Targeting Immune Checkpoints for the Treatment of Solid Tumors. *Antibodies (Basel, Switzerland)*, *8*(4), 51. http://doi.org/10.3390/antib8040051

Gu, B., Zhang, J., Chen, Q., Tao, B., Wang, W., Zhou, Y., et al. (2010). Aire regulates the expression of differentiation-associated genes and self-renewal of embryonic stem cells. *Biochemical and Biophysical Research Communications*, *394*(2), 418–423. http://doi.org/10.1016/j.bbrc.2010.03.042

Hanna, S., & Etzioni, A. (2014). MHC class I and II deficiencies. *The Journal of Allergy and Clinical Immunology*, *134*(2), 269–275. http://doi.org/10.1016/j.jaci.2014.06.001

Heinonen, S., Rodriguez-Fernandez, R., Diaz, A., Oliva Rodriguez-Pastor, S., Ramilo, O., & Mejias, A. (2019). Infant Immune Response to Respiratory Viral Infections.

*Immunology and Allergy Clinics of North America, 39*(3), 361–376. http://doi.org/10.1016/j.iac.2019.03.005

Henning, A. N., Klebanoff, C. A., & Restifo, N. P. (2018). Silencing stemness in T cell differentiation. *Science, 359*(6372), 163–164. http://doi.org/10.1126/science.aar5541

Hewitt, E. W. (2003). The MHC class I antigen presentation pathway: strategies for viral immune evasion. *Immunology, 110*(2), 163–169. http://doi.org/10.1046/j.1365-2567.2003.01738.x

Inglesfield, S., Cosway, E. J., Jenkinson, W. E., & Anderson, G. (2019). Rethinking Thymic Tolerance: Lessons from Mice. *Trends in Immunology, 40*(4), 279–291. http://doi.org/10.1016/j.it.2019.01.011

Janssen, B. J. C., Huizinga, E. G., Raaijmakers, H. C. A., Roos, A., Daha, M. R., Nilsson-Ekdahl, K., et al. (2005). Structures of complement component C3 provide insights into the function and evolution of immunity. *Nature, 437*(7058), 505–511. http://doi.org/10.1038/nature04005

Katkere, B., Rosa, S., Caballero, A., Repasky, E. A., & Drake, J. R. (2010). Physiological-range temperature changes modulate cognate antigen processing and presentation mediated by lipid raft-restricted ubiquitinated B cell receptor molecules. *Journal of Immunology (Baltimore, Md. : 1950), 185*(9), 5032–5039. http://doi.org/10.4049/jimmunol.1001653

Keir, M. E., Liang, S. C., Guleria, I., Latchman, Y. E., Qipo, A., Albacker, L. A., et al. (2006). Tissue expression of PD-L1 mediates peripheral T cell tolerance. *The Journal of Experimental Medicine, 203*(4), 883–895. http://doi.org/10.1084/jem.20051776

Kieback, E., Hilgenberg, E., Stervbo, U., Lampropoulou, V., Shen, P., Bunse, M., et al. (2016). Thymus-Derived Regulatory T Cells Are Positively Selected on Natural Self-Antigen through Cognate Interactions of High Functional Avidity. *Immunity, 44*(5), 1114–1126. http://doi.org/10.1016/j.immuni.2016.04.018

Kisielow, P. (2019). How does the immune system learn to distinguish between good and evil? The first definitive studies of T cell central tolerance and positive selection. *Immunogenetics, 71*(8-9), 513–518. http://doi.org/10.1007/s00251-019-01127-8

Klein, J. S., Gnanapragasam, P. N. P., Galimidi, R. P., Foglesong, C. P., West, A. P., & Bjorkman, P. J. (2009). Examination of the contributions of size and avidity to the neutralization mechanisms of the anti-HIV antibodies b12 and 4E10. *Proceedings of the National Academy of Sciences of the United States of America, 106*(18), 7385–7390. http://doi.org/10.1073/pnas.0811427106

Kont, V., Laan, M., Kisand, K., Merits, A., Scott, H. S., & Peterson, P. (2008). Modulation of Aire regulates the expression of tissue-restricted antigens. *Molecular Immunology*, *45*(1), 25–33. http://doi.org/10.1016/j.molimm.2007.05.014

Kubes, P., & Jenne, C. (2018). Immune Responses in the Liver. *Annual Review of Immunology*, *36*(1), 247–277. http://doi.org/10.1146/annurev-immunol-051116-052415

Kuka, M., & Iannacone, M. (2018). Viral subversion of B cell responses within secondary lymphoid organs. *Nature Reviews. Immunology*, *18*(4), 255–265. http://doi.org/10.1038/nri.2017.133

Kurosaki, T., Kometani, K., & Ise, W. (2015). Memory B cells. *Nature Reviews. Immunology*, *15*(3), 149–159. http://doi.org/10.1038/nri3802

la Peña, de, A. H., Goodall, E. A., Gates, S. N., Lander, G. C., & Martin, A. (2018). Substrate-engaged 26S proteasome structures reveal mechanisms for ATP-hydrolysis-driven translocation. *Science*, *362*(6418), eaav0725. http://doi.org/10.1126/science.aav0725

Le Nours, J., Shahine, A., & Gras, S. (2018). Molecular features of lipid-based antigen presentation by group 1 CD1 molecules. *Seminars in Cell & Developmental Biology*, *84*, 48–57. http://doi.org/10.1016/j.semcdb.2017.11.002

Li, X., Zai, J., Zhao, Q., Nie, Q., Li, Y., Foley, B. T., & Chaillon, A. (2020). Evolutionary history, potential intermediate animal host, and cross-species analyses of SARS-CoV-2. *Journal of Medical Virology*, *6*(6), 6–611. http://doi.org/10.1002/jmv.25731

Lindh, E., Lind, S. M., Lindmark, E., Hässler, S., Perheentupa, J., Peltonen, L., et al. (2008). AIRE regulates T-cell-independent B-cell responses through BAFF. *Proceedings of the National Academy of Sciences of the United States of America*, *105*(47), 18466–18471. http://doi.org/10.1073/pnas.0808205105

Liston, A., Lesage, S., Wilson, J., Peltonen, L., & Goodnow, C. C. (2003). Aire regulates negative selection of organ-specific T cells. *Nature Immunology*, *4*(4), 350–354. http://doi.org/10.1038/ni906

Longdon, B., Hadfield, J. D., Webster, C. L., Obbard, D. J., & Jiggins, F. M. (2011). Host phylogeny determines viral persistence and replication in novel hosts. *PLoS Pathogens*, *7*(9), e1002260. http://doi.org/10.1371/journal.ppat.1002260

Lu, C., Zanker, D., Lock, P., Jiang, X., Deng, J., Duan, M., et al. (2019). Memory regulatory T cells home to the lung and control influenza A virus infection. *Immunology and Cell Biology*, *97*(9), 774–786. http://doi.org/10.1111/imcb.12271

Ma, C. S., & Phan, T. G. (2017). Here, there and everywhere: T follicular helper cells on the move. *Immunology*, *152*(3), 382–387. http://doi.org/10.1111/imm.12793

Ma, C. S., Uzel, G., & Tangye, S. G. (2014). Human T follicular helper cells in primary immunodeficiencies. *Current Opinion in Pediatrics*, *26*(6), 720–726. http://doi.org/10.1097/MOP.0000000000000157

Mårtensson, I.-L., Almqvist, N., Grimsholm, O., & Bernardi, A. I. (2010). The pre-B cell receptor checkpoint. *FEBS Letters*, *584*(12), 2572–2579. http://doi.org/10.1016/j.febslet.2010.04.057

Menachery, V. D., Yount, B. L., Debbink, K., Agnihothram, S., Gralinski, L. E., Plante, J. A., et al. (2015). A SARS-like cluster of circulating bat coronaviruses shows potential for human emergence. *Nature Medicine*, *21*(12), 1508–1513. http://doi.org/10.1038/nm.3985

Menachery, V. D., Yount, B. L., Sims, A. C., Debbink, K., Agnihothram, S. S., Gralinski, L. E., et al. (2016). SARS-like WIV1-CoV poised for human emergence. *Proceedings of the National Academy of Sciences of the United States of America*, *113*(11), 3048–3053. http://doi.org/10.1073/pnas.1517719113

Methot, S. P., & Di Noia, J. M. (2017). Molecular Mechanisms of Somatic Hypermutation and Class Switch Recombination. *Advances in Immunology*, *133*, 37–87. http://doi.org/10.1016/bs.ai.2016.11.002

Mueller, S. N., Gebhardt, T., Carbone, F. R., & Heath, W. R. (2013). Memory T cell subsets, migration patterns, and tissue residence. *Annual Review of Immunology*, *31*(1), 137–161. http://doi.org/10.1146/annurev-immunol-032712-095954

Muro, R., Takayanagi, H., & Nitta, T. (2019). T cell receptor signaling for γδT cell development. *Inflammation and Regeneration*, *39*(1), 6–11. http://doi.org/10.1186/s41232-019-0095-z

Ng, J. H. J., Tachedjian, M., Deakin, J., Wynne, J. W., Cui, J., Haring, V., et al. (2016). Evolution and comparative analysis of the bat MHC-I region. *Scientific Reports*, *6*(1), 21256–18. http://doi.org/10.1038/srep21256

Nguyen, A., David, J. K., Maden, S. K., Wood, M. A., Weeder, B. R., Nellore, A., & Thompson, R. F. (2020). Human leukocyte antigen susceptibility map for SARS-CoV-2. *Journal of Virology*, *94*(13), 727. http://doi.org/10.1128/JVI.00510-20

Nicolai, S., Wegrecki, M., Cheng, T.-Y., Bourgeois, E. A., Cotton, R. N., Mayfield, J. A., et al. (2020). Human T cell response to CD1a and contact dermatitis allergens in botanical extracts and commercial skin care products. *Science Immunology*, *5*(43), eaax5430. http://doi.org/10.1126/sciimmunol.aax5430

Nutt, S. L., Hodgkin, P. D., Tarlinton, D. M., & Corcoran, L. M. (2015). The generation of antibody-secreting plasma cells. *Nature Reviews. Immunology*, *15*(3), 160–171. http://doi.org/10.1038/nri3795

O'Neill, L. A. J., Kishton, R. J., & Rathmell, J. (2016). A guide to immunometabolism for immunologists. *Nature Reviews. Immunology*, *16*(9), 553–565. http://doi.org/10.1038/nri.2016.70

Olson, W. J., Jakic, B., & Hermann-Kleiter, N. (2020). Regulation of the Germinal Center Response by Nuclear Receptors and Implications for Autoimmune Diseases. *The FEBS Journal*, *41*, 529. http://doi.org/10.1111/febs.15312

Pavlovich, S. S., Lovett, S. P., Koroleva, G., Guito, J. C., Arnold, C. E., Nagle, E. R., et al. (2018). The Egyptian Rousette Genome Reveals Unexpected Features of Bat Antiviral Immunity. *Cell*, *173*(5), 1098–1102.e18. http://doi.org/10.1016/j.cell.2018.03.070

Poggio, M., Hu, T., Pai, C.-C., Chu, B., Belair, C. D., Chang, A., et al. (2019). Suppression of Exosomal PD-L1 Induces Systemic Anti-tumor Immunity and Memory. *Cell*, *177*(2), 414–427.e13. http://doi.org/10.1016/j.cell.2019.02.016

Poli, A., Michel, T., Patil, N., & Zimmer, J. (2018). Revisiting the Functional Impact of NK Cells. *Trends in Immunology*, *39*(6), 460–472. http://doi.org/10.1016/j.it.2018.01.011

Pradeu, T., & Pasquier, Du, L. (2018). Immunological memory: What's in a name? *Immunological Reviews*, *283*(1), 7–20. http://doi.org/10.1111/imr.12652

Praest, P., Liaci, A. M., Förster, F., & Wiertz, E. J. H. J. (2019). New insights into the structure of the MHC class I peptide-loading complex and mechanisms of TAP inhibition by viral immune evasion proteins. *Molecular Immunology*, *113*, 103–114. http://doi.org/10.1016/j.molimm.2018.03.020

Próchnicki, T., & Latz, E. (2017). Inflammasomes on the Crossroads of Innate Immune Recognition and Metabolic Control. *Cell Metabolism*, *26*(1), 71–93. http://doi.org/10.1016/j.cmet.2017.06.018

Pupovac, A., & Good-Jacobson, K. L. (2017). An antigen to remember: regulation of B cell memory in health and disease. *Current Opinion in Immunology*, *45*, 89–96. http://doi.org/10.1016/j.coi.2017.03.004

Rankin, L. C., & Artis, D. (2018). Beyond Host Defense: Emerging Functions of the Immune System in Regulating Complex Tissue Physiology. *Cell*, *173*(3), 554–567. http://doi.org/10.1016/j.cell.2018.03.013

Samir, P., & Kanneganti, T.-D. (2019). Hidden Aspects of Valency in Immune System Regulation. *Trends in Immunology*, *40*(12), 1082–1094. http://doi.org/10.1016/j.it.2019.10.005

Schildberg, F. A., Klein, S. R., Freeman, G. J., & Sharpe, A. H. (2016). Coinhibitory Pathways in the B7-CD28 Ligand-Receptor Family. *Immunity*, *44*(5), 955–972. http://doi.org/10.1016/j.immuni.2016.05.002

Schoggins, J. W. (2019). Interferon-Stimulated Genes: What Do They All Do? *Annual Review of Virology*, *6*(1), 567–584. http://doi.org/10.1146/annurev-virology-092818-015756

Schoggins, J. W., & Rice, C. M. (2011). Interferon-stimulated genes and their antiviral effector functions. *Current Opinion in Virology*, *1*(6), 519–525. http://doi.org/10.1016/j.coviro.2011.10.008

Schreiner, D., & King, C. G. (2018). CD4+ Memory T Cells at Home in the Tissue: Mechanisms for Health and Disease. *Frontiers in Immunology*, *9*, 2394. http://doi.org/10.3389/fimmu.2018.02394

Sharpe, A. H., & Pauken, K. E. (2018). The diverse functions of the PD1 inhibitory pathway. *Nature Reviews. Immunology*, *18*(3), 153–167. http://doi.org/10.1038/nri.2017.108

Shortman, K., & Heath, W. R. (2010). The CD8+ dendritic cell subset. *Immunological Reviews*, *234*(1), 18–31. http://doi.org/10.1111/j.0105-2896.2009.00870.x

Srivastava, S., Grace, P. S., & Ernst, J. D. (2016). Antigen Export Reduces Antigen Presentation and Limits T Cell Control of M. tuberculosis. *Cell Host & Microbe*, *19*(1), 44–54. http://doi.org/10.1016/j.chom.2015.12.003

Stebegg, M., Kumar, S. D., Silva-Cayetano, A., Fonseca, V. R., Linterman, M. A., & Graca, L. (2018). Regulation of the Germinal Center Response. *Frontiers in Immunology*, *9*, 2469. http://doi.org/10.3389/fimmu.2018.02469

tenOever, B. R. (2016). The Evolution of Antiviral Defense Systems. *Cell Host & Microbe*, *19*(2), 142–149. http://doi.org/10.1016/j.chom.2016.01.006

Theisen, D., & Murphy, K. (2017). The role of cDC1s in vivo: CD8 T cell priming through cross-presentation. *F1000Research*, 6(98), 98. http://doi.org/10.12688/f1000research.9997.1

Tom, J. K., Albin, T. J., Manna, S., Moser, B. A., Steinhardt, R. C., & Esser-Kahn, A. P. (2019). Applications of Immunomodulatory Immune Synergies to Adjuvant Discovery and Vaccine Development. *Trends in Biotechnology*, *37*(4), 373–388. http://doi.org/10.1016/j.tibtech.2018.10.004

Tuttle, K. D., Krovi, S. H., Zhang, J., Bedel, R., Harmacek, L., Peterson, L. K., et al. (2018). TCR signal strength controls thymic differentiation of iNKT cell subsets. *Nature Communications*, *9*(1), 2650–13. http://doi.org/10.1038/s41467-018-05026-6

Vabret, N., Britton, G. J., Gruber, C., Hegde, S., Kim, J., Kuksin, M., et al. (2020). Immunology of COVID-19: Current State of the Science. *Immunity*, *52*(6), 910–941. http://doi.org/10.1016/j.immuni.2020.05.002

van de Weijer, M. L., Luteijn, R. D., & Wiertz, E. J. H. J. (2015). Viral immune evasion: Lessons in MHC class I antigen presentation. *Seminars in Immunology*, *27*(2), 125–137. http://doi.org/10.1016/j.smim.2015.03.010

Walker, J. A., & McKenzie, A. N. J. (2018). TH2 cell development and function. *Nature Reviews. Immunology*, *18*(2), 121–133. http://doi.org/10.1038/nri.2017.118

Wang, N., Li, S.-Y., Yang, X.-L., Huang, H.-M., Zhang, Y.-J., Guo, H., et al. (2018). Serological Evidence of Bat SARS-Related Coronavirus Infection in Humans, China. *Virologica Sinica*, *33*(1), 104–107. http://doi.org/10.1007/s12250-018-0012-7

Ward-Kavanagh, L. K., Lin, W. W., Šedý, J. R., & Ware, C. F. (2016). The TNF Receptor Superfamily in Co-stimulating and Co-inhibitory Responses. *Immunity*, *44*(5), 1005–1019. http://doi.org/10.1016/j.immuni.2016.04.019

West, E. E., Kolev, M., & Kemper, C. (2018). Complement and the Regulation of T Cell Responses. *Annual Review of Immunology*, *36*(1), 309–338. http://doi.org/10.1146/annurev-immunol-042617-053245

Wood, K. J., Bushell, A. R., & Jones, N. D. (2010). The discovery of immunological tolerance: now more than just a laboratory solution. *Journal of Immunology (Baltimore, Md. : 1950)*, *184*(1), 3–4. http://doi.org/10.4049/jimmunol.0990108

Woolhouse, M. E. J., Haydon, D. T., & Antia, R. (2005). Emerging pathogens: the epidemiology and evolution of species jumps. *Trends in Ecology & Evolution*, *20*(5), 238–244. http://doi.org/10.1016/j.tree.2005.02.009

Wu, Aiping, Peng, Y., Huang, B., Ding, X., Wang, X., Niu, P., et al. (2020a). Genome Composition and Divergence of the Novel Coronavirus (2019-nCoV) Originating in China. *Cell Host & Microbe*, *27*(3), 325–328. http://doi.org/10.1016/j.chom.2020.02.001

Wu, Yan, Wang, F., Shen, C., Peng, W., Li, D., Zhao, C., et al. (2020b). A noncompeting pair of human neutralizing antibodies block COVID-19 virus binding to its receptor ACE2. *Science*, *368*(6496), 1274–1278. http://doi.org/10.1126/science.abc2241

Wu, Yanling, Li, C., Xia, S., Tian, X., Kong, Y., Wang, Z., et al. (2020c). Identification of Human Single-Domain Antibodies against SARS-CoV-2. *Cell Host & Microbe*, *27*(6), 891–898.e5. http://doi.org/10.1016/j.chom.2020.04.023

Xie, F., Xu, M., Lu, J., Mao, L., & Wang, S. (2019). The role of exosomal PD-L1 in tumor progression and immunotherapy. *Molecular Cancer*, *18*(1), 146–10. http://doi.org/10.1186/s12943-019-1074-3

Yamazaki, T., Akiba, H., Iwai, H., Matsuda, H., Aoki, M., Tanno, Y., et al. (2002). Expression of programmed death 1 ligands by murine T cells and APC. *Journal of*

*Immunology (Baltimore, Md. : 1950)*, *169*(10), 5538–5545. http://doi.org/10.4049/jimmunol.169.10.5538

Yu, D., & Ye, L. (2018). A Portrait of CXCR5+ Follicular Cytotoxic CD8+ T cells. *Trends in Immunology*, *39*(12), 965–979. http://doi.org/10.1016/j.it.2018.10.002

Zhang, G., Cowled, C., Shi, Z., Huang, Z., Bishop-Lilly, K. A., Fang, X., et al. (2013). Comparative analysis of bat genomes provides insight into the evolution of flight and immunity. *Science*, *339*(6118), 456–460. http://doi.org/10.1126/science.1230835

Zhang, Y.-Z., & Holmes, E. C. (2020). A Genomic Perspective on the Origin and Emergence of SARS-CoV-2. *Cell*, *181*(2), 223–227. http://doi.org/10.1016/j.cell.2020.03.035

Zhou, Z., He, H., Wang, K., Shi, X., Wang, Y., Su, Y., et al. (2020). Granzyme A from cytotoxic lymphocytes cleaves GSDMB to trigger pyroptosis in target cells. *Science*, *146*(6494), eaaz7548. http://doi.org/10.1126/science.aaz7548

Zhuang, X., & Long, E. O. (2019). Inhibition-Resistant CARs for NK Cell Cancer Immunotherapy. *Trends in Immunology*, *40*(12), 1078–1081. http://doi.org/10.1016/j.it.2019.10.004

Manufactured by Amazon.ca
Bolton, ON